Designing a National Scientific and Technological Communication System

RUSSELL L. ACKOFF | MARTIN C. J. ELTON
THOMAS A. COWAN | JAMES C. EMERY
PETER DAVIS | MARYBETH L. MEDITZ
WLADIMER M. SACHS

With the collaboration of

WILLIAM C. DONOVAN
MOHAMMAD JAHANI
JONATHAN V. ROSENHEAD

Designing a National Scientific and Technological Communication System

University of Pennsylvania Press/1976

In memory of

MARVIN REES

CONTENTS

ILLUSTRATIONS

TABLES

GLOSSARY
OF ACRONYMS

ACM—Association for Computing Machinery
APA—American Psychological Association
APTIC—EPA information system
ARPANET—Advanced Research Projects Agency Network
ASCA—Automatic Subject Citation Alert
ASIS—American Society for Information Science
BASIC—Biological Abstracts Subjects in Context
BIOSIS—Biosciences Information Service
CA—Chemical Abstracts
CAIN—Cataloging And INdexing service of NAL
CAN/SDI—Canadian Selective Dissemination
        of Information
CAS—Chemical Abstract Service
CAST—Center for Advancement of Science and Technology
        (proposed for NSF)
CLASS—Current Literature Altering Search Service
CPU—Central Processing Unit
CRT—Cathode Ray Tube
DIALOG—information retrieval system of Lockheed
DoD—Department of Defense
EDUCOM—Interuniversity Communications Council
EPA—Environmental Protection Agency
ERIC—Education Research Information Center
        (U.S. Office of Education)
ERSO/ECDO—space document service
ESR—Entry, Storage, Retrieval

GTE—General Telephone and Electronics

IBM—International Business Machines

IEEE—Institute of Electrical and Electronic Engineers

ILTRI—Illinois Institute of Technology Research Institute

INFROSS—Investigation into Information Requirements
of the Social Sciences (Bath University)

INQUIRE—software package provided by
Infodata Systems

ISI—Institute for Scientific Information

ISS—International SCATT System

ITRIC—IBM, Technical Information Retrieval Center

KWAC—Keyword-And-Context

KWIC—Keyword-In-Context

KWOC—Keyword-Out-of-Context

LSC—Local SCATT Center

MARC—MAchine-Readable Catalog (Library of Congress)

MEDLARS—MEDical Literature Analysis and
Retrieval System (NLM)

MEDLINE—MEDLARS on-LINE (NLM)

MIC—Mechanized Information Center, information center service at
Ohio State University

MIT—Massachusetts Institute of Technology

NAL—National Agriculture Library

NASA—National Aeronautics and Space Administration

NFAIS—National Federation of Abstracting and
Indexing Services

NFSAIS—National Federation of Science Abstracting and
Indexing Services

NINDS—National Institute of Neurological Disease
and Stroke

NLM—National Library of Medicine

NSC—National SCATT Center

NSF—National Science Foundation

NSS—National SCATT System

OATS—Original Article Tearsheet Service

OECD—Organization for Economic Cooperation and Development

ORBIT—on-line information retrieval system of Systems
Development Corporation

OSIS—Office of Science Information Service (NSF)

PESTDOC—database of the RINGDOC system

R&D—Research and Development

RECON—Retrospective Conversion of Bibliographic Records
(Library of Congress)

RINGDOC—document distribution system developed by Pfizer
Pharmaceuticals and offered by Derwent Publications, Ltd.

RSC—Regional SCATT Center
S—Similarity Index
SAE—Society for Automotive Engineers
SATCOM—Committee on Scientific and Technical Communication
(National Academy of Sciences–National Academy of Engineers)
SCAN—Selected Current Aerospace Notices (NASA)
SCATT—Scientific Communication and Technology Transfer
SDC—Systems Development Corporation
SDI—Selective Dissemination of Information
UCLA—University of California at Los Angeles
VETDOC—database of the RINGDOC system

FOREWORD

Decentralization is the most striking characteristic of the scientific and technical information enterprise in the United States. Numerous overlapping and often competitive services are offered by professional societies and other not-for-profit organizations, profit-seeking firms, and federal agencies. Each has developed with its own rationale and has sought its place among the welter of existing information services. In the past, this system worked reasonably well—when each system served a particular group of users whose interests generally coincided with the literature covered by the system. Today, however, conditions have changed. Research on major national problems requires ready access to information drawn from many disciplines. Now, unnecessary incompatibilities and redundancies among systems add to confusion among users and increase overall costs in acquiring available information.

Users, purchasers, and sellers of information services equally lament present obstacles to acquiring and using information. We know we cannot go back to earlier, simpler days. Also, we realize we cannot force needed improvements on users, buyers, and sellers of information services. But we *can* attempt to formulate and encourage the adoption of an ideal system that combines the best of all possible arrangements for providing access to available information. This offers a way to balance our democratic respect for pluralism with our managerial requirements for efficiency and effectiveness.

Dr. Russell Ackoff has pioneered in studies aimed at developing ideal systems for a number of problems. Now, with support from the National Science Foundation's Office of Science Information Service, Dr. Ackoff and his research team have begun developing an ideal system for the scientific and technical information enterprise in the United States. This development requires successive revision of a conceptual framework for organizing the flow of information from points of origin to all possible points of application. One of the beauties of this approach is that all affected parties—information generators and users as well as information processors—can help shape the evolving model. Consequently, as outlined in this book, the ideal system for scientific and technical information, as seen by Dr. Ackoff and his colleagues, is not *the* system to be implemented, but represents, instead, a first approximation of what might be possible, within our decentralized, pluralistic scientific and technical community. Implementation of any elements of the proposed ideal system of course rests upon voluntary cooperative arrangements among the affected parties.

The Office of Science Information sees a number of benefits from development of an ideal system design. With the outline presented in this book, we have a basis for structured interchanges among all parties involved in scientific and technical information exchange and technology transfer efforts. Stakeholders in these activities can have a voice in shaping national directions. Also, for the first time, we have an explicit alternative to our usual pattern of incremental drift in the evolution of national capabilities of transferring information.

As we celebrate our Bicentennial, Dr. Ackoff offers the information and technology transfer communities and the persons they serve an opportunity to join in a national town meeting to debate and help shape the framework for services in the coming decades. I hope readers, especially researchers and users, will join information processors in active development of the evolving ideal system design. Send comments directly to Dr. Ackoff. You may want to critique individual elements of the proposed design or the system as a whole. But do contribute. Together we can clarify options and be more assured that changes we make will result in greater long-term benefits.

Lee G. Burchinal, Head
Office of Science Information Service
National Science Foundation

PROLOGUE

The primary purpose of the project reported here is not what it appears to be: to produce a design of a National Scientific Communication and Technology Transfer (SCATT) System that would in some sense be preferable to the one currently available. Rather, its purpose is to mobilize the large number of relatively autonomous subsystems of the current system into a collaborative effort directed at redesigning their system and implementing their design. The idealized design process used is intended to stimulate such a mobilization, and the design produced is intended to serve as a platform from which that mobilization can spring.

For reasons that are discussed in Chapter 1 the idealized design process has no end point. Therefore, this report was prepared at an arbitrary point in time: at the end of the grant that made our work possible. Fortunately, however, our work has been continued into a second phase through the continued support of the National Science Foundation. In the second phase we are attempting to involve as many parts of the system as we can in the design process and to make the transition from idealized design to practical planning. Such planning is being directed at determining the extent to which a design such as is presented here can be approximated in reality.

This report is the product of the sixth iteration of the design process described in Chapter 1. Its product, like those of previous iterations, is controversial. This is bound to be the case in any effort that addresses itself to important issues and that does so in an unconventional way. Our work has attracted strong opposition as well as enthusiastic support. Fortunately for us, support appears to outweigh opposition.

Our intention has not been to impose our design of a National SCATT

1

System upon the members of the scientific and technological community, but rather to develop a design that this community would voluntarily take steps to adopt. Since we believe that perfection is an ideal that can never be attained, we have designed a system that is flexible, adaptive, and easy to change. We have been encouraged by the responses to the potential for progressive development that we attempted to design into the system. One of our principal advisors, Professor Jordan J. Baruch of Dartmouth College (personal communication, June 30, 1975), expressed this aspiration much better than we could:

> Unlike the fields of physics and chemistry, information science has been deterred in its development by the lack of an appropriate environment in which to perform realistic experiments. In chemistry and physics, the laboratory environment is both highly controlled and a realistic simulation of the external world is possible. The control aspect makes analysis and deduction possible while the reality makes extrapolation to real world activity a reasonable step. In information science, control has been achieved almost always at the sacrifice of reality while realistic experiments, by and large, had to be conducted in a relatively uncontrolled environment. As a result, information science has grown along two paths, the highly theoretical and the practical albeit largely descriptive.
>
> Regardless of whatever else the proposed "Idealized System" accomplishes, it will provide the first laboratory in information science to combine the requirements of both control and quality. Since the proposed system permits operating with any set of constraints imposed by the experimenter on a target population, economic variables, linguistic variables, modality variables, etc. can all be explored simultaneously.
>
> Significant questions such as the economic value of reader feedback via marginalia can be explored as easily as the importance of response time. Indeed, if the system is viewed as a national information science laboratory resource, if it is designed so that experimental constraints can be imposed on authorized user subsets from a remote location and if it provides an internal facility for guaranteeing the commensurability of data, it will both permit and encourage a level of experimentation and of experimental sophistication not now available.

These observations by Professor Baruch relate to a second objective of this effort: to provide the Office of Scientific Information Service of the National Science Foundation with a basis for encouraging and selecting research and development projects that would facilitate and accelerate improvement of relevant services. It is difficult to assess the importance of independently conceived research and development proposals without some concept of what one is trying to do. Therefore, our effort has been directed toward providing this Office with such a concept, a concept collaboratively produced by a significant number of those who have dedicated themselves to improvement of scientific and technological information services.

Although many have participated in the design process the product of which is presented here, responsibility for this product is ours, whatever its deficiencies. We have had to mediate between many incompatible suggestions and resolve numerous conflicts of opinion. Such incompatibilities and conflicts arose not only among our collaborators but also within the team that worked on this project. Had the members of the team been in agreement on each internally and externally generated issue, the design would not have been so rich or so provocative as we believe it to be.

Perhaps no issue has generated so much difference of opinion as the organization of this report. We could find no ordering of its parts that would satisfy all. Therefore, the reader should feel free to wander through the report however he or she sees fit.

In Chapter 1 we describe the nature of idealized design as we have developed it. Chapter 2 is a description of our Idealized SCATT System. Chapter 3 is a more detailed account of the processes involved in the generation, storage, and retrieval of scientific and technological information. Chapter 4 deals with the technological and hardware aspects of the System, and shows, we believe, that the design is technologically feasible. Chapter 5 deals with the financial and economic aspects of the System and is aimed at establishing its operational viability. In Chapter 6 we show how the national model could be expanded into an international network of such systems. Chapter 7 provides a description of currently available services and the means by which they are provided. Though much of this material is familiar to specialists, we believe that many readers will find it both helpful and revealing. It provides a description of the current state of the art and science of scientific communication and technology transfer. It does not purport to be definitive or all inclusive, but since it served as a starting point in our design process, we feel it might be useful to others.

Finally, in the Epilogue we do homage to our many critics and advisors by trying to answer some of the profound philosophical and methodological questions they have raised for us. We are especially grateful to Dr. Joel Goldhar and Dr. Carole Ganz of the Office of Science Information Service of the National Science Foundation for their unwavering support of this exciting pioneering venture.

We hope that the reader will have as much fun in reading this report as we have had in producing it. May he have reward enough to justify his effort and ours.

1

IDEALIZATION
The Nature of Idealized Design

The study reported here is based on a process we call "idealized design," which is part of the "whole-system" approach to planning. In this chapter we introduce first the need for the whole-system approach in the field of scientific communication and technology transfer. Second, we show how idealized design is related to the more general process of planning.

THE NEED FOR THE WHOLE-SYSTEM APPROACH

The Scientific Communication and Technology Transfer (SCATT) "System," if it may be called that, currently operating in the United States consists of a large complex of interacting processes and institutions that design, produce, distribute, and market information produced by science and technology. This complex is not an organized system. It is an aggregate of independently controlled parts whose activities, in general, are neither integrated nor coordinated.

Numerous studies have been directed at improving the efficiency of each part of the current system independently of the others. The cumulative effect of these studies on the effectiveness of the system has been minimal. Users still complain about a growing information overload, a lack of effective filtration and condensation of messages, an increasing lead time to publication, retrieval problems, and so on and on. This state of affairs is not surprising because improvement of the efficiency of each part of any system taken separately seldom produces an improvement of that system's overall performance. For example, if we were to select the best automobile transmission, generator, carburetor, frame, and so on, regardless of make, and try to assemble them we would not obtain a well-functioning automobile. In fact, it is very unlikely that it would function at all because the parts would not fit together well.

A system's performance depends more on the interaction of its parts than on their actions considered independently of each other. This is true even of highly integrated and coordinated systems, all of whose parts are under one management. It is even truer of a pluralistic system whose parts are virtually autonomous and therefore can be "systematized" only through a combination of public policies and voluntary collaboration.

When the redesign of one part of a system is undertaken independently of the redesign of other parts, the range of possibilities that are considered to be feasible is severely limited. For example, the variety of possible changes of the living room of a house, if one assumes that no other part of the house is to be changed, is much more constrained than it would be if remodelling—and certainly if reconstruction—of the entire house were considered. The changes one would contemplate in technical libraries, to take another example, would be severely constrained if one assumed that no changes were to be made in the nature or form of the documentation which they received, stored, and distributed.

For these reasons there are great benefits to be derived from considering the redesign of a SCATT (or any) system as a whole. By examining combinations of changes in the parts, one can contemplate larger potential effects on the whole. New possibilities are uncovered for both parts and the whole. The focus is appropriately changed: the characteristics of the whole are not viewed as a resultant of the characteristics of the parts; rather, the characteristics of the parts are derived from desirable characteristics of the whole.

Even designs of and plans for a system as a whole can be severely constrained by restrictions that designers and planners incorrectly assume are imposed on them. As we will argue below, at least some of the self-imposed constraints on the creativity of systems designers can be lifted by engaging them in the idealized (re)design process.

THE NATURE OF IDEALIZATION[1]

Idealization is a design process subject to only two constraints:

(1)   *The design may not involve any technology that is not now known to be feasible.*

This constraint does not preclude extensions of current technology or technological innovation; it merely limits design to technologies known to

---

[1] For a more extensive discussion of idealization and examples of its application, see Ackoff (1974).

be possible. For example, the idealizer may assume the availability of picture telephone or even facsimile color transmission into the home because these technologies exist. On the other hand, he cannot assume the availability of an antigravity device or pervasive mental telepathy. In short, an idealized design is not a work of science fiction; it is a feasible, however improbable, design.

(2) *The system designed must be operationally viable. It must be capable of operating if it did come into existence.*

The judgment as to whether or not the design produced is both technologically feasible and operationally viable should be made by those who hold a stake in the system; in particular, by those who participate in the design process. However, their judgment should be subjected to critical review by persons who are familiar with both the system and the technology involved.

An idealization is a design of the system that the designers would create now if they could create any technologically feasible and operationally viable system they wanted. Any design they produce is necessarily incomplete and imperfect because there are always some relevant questions which do not occur to the designers and others to which they do not have ideal answers; but the designers do not have to resolve every question in the design they produce. Where they cannot answer a question they can design into the system a capability of finding the right answer. For example, if, in designing a community, they cannot decide whether they prefer a tax on income or on consumption, they can incorporate into their design of the system either an experiment that will provide an answer, or a choice between the alternatives so that the system's participants can make the decision. Furthermore, since any system and its environment will change in ways that its designers cannot anticipate, the system should be designed to learn effectively from its own experience and to adapt to environmental changes. This requires that it be flexible and easy to change.

The product of an idealized design is not a design of an ideal system, rather, it is an ideal-seeking system. Such a system is neither static nor perfect, but is one that is capable of approaching its designers' conception of perfection. This requires that the system be designed to have the capability of learning and modifying itself rapidly and effectively. It should also lend itself to, and facilitate, continuous experimentation that is directed to improving its performance.

Because the product of an idealized design is an ideal-seeking, not an ideal, system, it is not utopian. Furthermore, idealized-system designers recognize that their concept of the ideal will change with time. With more experience and thought in either the real or idealized world, they would improve their design.[2] Therefore, an idealized design is a relative absolute. It is absolute in the sense that it is a representation of the designers' current conception of their relevant ultimate values. It is relative in the sense that it is a representation involving the designers' currently imperfect information about, and knowledge and understanding of, the system being designed—not to mention their less than perfect wisdom.

If an idealized design is always both incomplete and imperfect, why bother to produce it? Our experience with such design in both the public and private domains reveals that there are at least six important reasons for doing so. Consider each in turn:

(1) *The idealized-design process converts planning from a retrospective to a prospective orientation.*

Most system redesign and planning is reactive—preoccupied with identifying and removing deficiencies in the current performance of system components. For example, librarians add clerks to accelerate availability of acquisitions, journals increase their length or frequency of publication to reduce backlogs of accepted but as yet unpublished articles, and new journals are created to fill unfilled needs. Reactive design and planning is an effort to move *from* what one does not want rather than toward what one does.

One who walks into the future facing the past has no control over where he is going. It may well be that in an idealized SCATT System we would not have journals as currently conceived. If this were the case, it would be clear that the current proliferation of journals makes progress toward such a state more difficult.

(2) *Idealization invites and facilitates widespread participation of all the stakeholders in the system, all those who are potentially affected by it.*

The process of idealization requires no special skills; anyone can participate. It is usually fun and provides those engaged in it with a chance to think deeply about a system they care for. It enables them to become conscious of, and to express, their aspirations for themselves and for the system involved.

Not all those who have a stake in the system comprehend it equally well or in equal detail, but participants in idealized design need only involve themselves with those aspects of the system in which they are interested. Others interacting with them can help make them conscious of the potential effect of their design of a part of the system on other parts or on the system as a whole. Therefore, the idealized-design process

---

[2] An interesting example of what might be considered an early idealized design is D. J. Urquhart's (1948) proposal to the Royal Society in London for a distribution system for scientific and technological information.

enables those involved in it to increase their comprehension and appreciation of the system as a whole.

The aspirations of all stakeholders (who are explicitly identified and discussed in Chapter 2) are relevant in the idealized-design process because the system has a responsibility for serving their interests. By participating they can increase the chances that their interests will be well served.

(3) *Idealization tends to generate consensus among those who would otherwise disagree on what should be done about the system.*

Idealization involves specification of ultimate values. In our experience, people tend to agree more on such values than on ones that are more immediate or on the means by which they may be obtained. If the design process begins with considerations of means without making objectives, let alone ideals, explicit, disagreements tend to be generated. Such differences of opinion, in turn, tend to be generalized into more extensive disagreement or even hostility. However, if agreement is reached on ultimate objectives, disagreements over means or short-run goals are more easily resolved because those involved recognize a fundamental agreement that provides a basis for constructive resolution of apparent differences.

In a number of idealization exercises in which we have participated, stakeholders who seldom agree on the way to correct a current deficiency tend to agree on what the ultimate objectives of the system involved should be. It can be argued that they do so because ultimate objectives lack reality and therefore agreement on them does not require any real commitment. This argument ignores the intensity of man's commitment to ideals. Such commitments have driven man to violent revolutions and to nonviolent crusades for peace. Idealized design is a way of producing strong commitments that make possible fundamental transformations of a system. It is generally so perceived by those engaged in it.

On the other hand, deep-seated disagreements about ultimate objectives indicate either that the system itself is in deep difficulties or else (and this often happens) that the dissidents have no real stake in the enterprise and are willing to wreck it for the sake of argument. Idealization brings deep-seated difficulties to the surface where they can be more easily dealt with. Dissidents with no real stake in the enterprise tend to drop out.

(4) *Idealization encourages those involved to look at the whole system, rather than to focus on one or more parts to the exclusion of the others.*

As we have previously noted, when one is preoccupied with correcting deficiencies in a system or with the means by which short-range goals are to be obtained, one almost always focuses on those parts of the system that are believed to be responsible for the deficiency and ignores the others. But when one tries to produce an idealized design it is difficult to consider anything but the whole. For example, in normal transportation planning it is usual to separate motor, rail, air, and sea travel; to plan for each separately; and then to try to bring them together into a comprehensive plan. But in idealized design of a transportation system attention is focused on the interactions between existing modes and their possible interactions with new modes. Therefore, such design encourages a more holistic view of the system than is normally taken.

(5) *Idealization induces more creativity in design than would otherwise occur.*

Self-imposed constraints are the principal obstruction to creativity. These constraints are seldom conscious; when they are, they are not often believed to be self-imposed but are usually attributed to external sources. Because idealization is an unconstrained design process and because it is associated with the ideal rather than the real, those involved in it are more likely to give their imagination free rein than they usually do. In particular, since they are not obliged to produce what they conceive to be a feasible design, they are less likely to impose constraints on themselves that are derived from what they incorrectly believe others will or will not do. In idealization each designer is freed of any need to concern himself with how others will respond to his design.

Interacting participants in the idealized-design process drive each other to abandoning their preconceptions and self-imposed constraints, and to thinking and proposing "way out" designs. Ingenuity and imagination bring appreciation and admiration from others, not conformity. When the stakeholders who participate in the design process fail to free each other's imagination and thus to stimulate creativity, the professional planners who coordinate the process can and should intervene. The design presented in this report is intended to be just such an "intervention." Hopefully, it is not the end but the beginning of an expanding design effort.

(6) *Idealization enlarges the designers' concept of what is feasible.*

A system always has some properties that none of its parts do. An animal can run but no part of its body can. A less well-known corollary of this principle is this: a set of interrelated proposals, no one of which appears to be feasible when considered separately, may nevertheless be feasible when considered as a whole.

One of the most common experiences of those who have participated in an idealized design is the realization that it is more feasible than they expected. The principal obstructions to its realization, if there are any, seldom turn out to be economic, legal, or, in general, external factors, but the reluctance of people individually and collectively to change.

When those who believe an idealized design to be infeasible are asked to make changes that will make it feasible, the resulting design is almost always closer to the ideal than is one which is based on piecemeal corrections of the current system. An idealized design tends to create the desire to do better than one normally thinks is the best that can be done.

PLANNING[3]

Planning, as we conceive it, is the design of a desirable future and the invention and selection of ways of bringing it about. Idealized design is an essential part of planning. Therefore, since a part of a system cannot be thoroughly understood without an understanding of the system of which it is part, we turn to a discussion of our concept of planning. We begin by considering four basic principles for organizing planning efforts. These are participation, coordination, integration, and continuity.

Principles for Organizing the Planning Effort

Participation. The principal value of planning is *not* derived from consuming its product, a plan, but from participating in its production. Therefore, participation is planning's most important product. Involvement in this process changes those involved—their relevant information, knowledge, and understanding increase. As the process continues they often begin to move the system being planned for in directions indicated by the planning process long before a formal plan is produced.

It follows that to maximize the benefits to be obtained from planning, no social system should be planned for by either an outside organization or only a part of the system itself; the community as a whole should plan for itself. Those affected by, but not involved in, planning fail to share its principal benefit and, therefore, are usually reluctant to accept its product. Therefore, participation facilitates implementation of the plan as well as development of the participants.

The proper function of the professional planner is not to plan for others, but to help them plan effectively and efficiently for themselves. He should have responsibility for involving the system's stakeholders in its planning and for educating them in the methodology and content of planning. He should also provide them with or help them to acquire the information, knowledge, and understanding they need to plan well. In

---

[3] For a more extensive discussion of the planning process and the role of idealization in it, see Ackoff (1970).

short, the professional planner should be an educator; he cannot learn for his students but he can and must motivate and facilitate the students' learning for themselves.

Coordination and Integration. Large systems are organized both horizontally and vertically. They are divided vertically into different levels; for example, national, state, and local. The higher the level, the more general is its responsibility. The activities at each level are divided (horizontally) usually either by functions (e.g., production and distribution), by type of product (e.g., books, articles, and patents), by geography (e.g., regions, states, or counties), or by some combination of these. Different levels of the same system may be divided either similarly or dissimilarly.

The effort to control the interactions of different activities at the same level of a system is coordinative; the effort to control the interactions of different levels is integrative. Planning itself should be both coordinated and integrated.

We have already made the general point that no part of a system can be planned for as effectively as is possible if it is planned for independently of other parts of the system. This is particularly true of complex systems, such as SCATT, planning for which is customarily carried out separately for each function and often for each type of product and geographical area. Such lack of coordination frequently precludes finding the best way of treating a deficiency that appears in one part of the system.

What appears to be a problem in one part of a system is often a symptom of a deficiency either of another part or of the way parts interact. For this reason the performance of a deficient part of a system is often subject to greater improvement by changing the behavior of other parts than by changing the behavior of the part that seems deficient. The opportunities for such improvement are likely to be overlooked in planning unless it is coordinated; that is, unless planning for each part interacts with planning for every other part.

It is often argued that because resources and time for planning are limited, only part or parts of the system can be planned for at any one time. Therefore, priorities are set and planning is focused on what are considered to be the most critical parts. Such "crisis planning" often fails to remove (although it may subdue) the crisis and even more often it fails to improve the system's overall performance. In planning, breadth is more important than depth: coverage is more important than detail. This is not to deny the value of depth or detail, but to assert that breadth or coverage should not be sacrificed for them.

Different types of decisions tend to be made at different levels of a system. *Operational* decisions, selection of means, are usually made at the bottom; *tactical* decisions, selection of short-range goals, at the next higher level; *strategic* decisions, selection of intermediate range objectives, at a still higher level; and *normative* decisions, selection of long-range and ultimate objectives, at the highest level.

Most planning flows either from the top down, thereby emphasizing only normative and strategic decisions; or from the bottom up, emphasizing only tactical and operating decisions. In our view, all such decisions should receive equal emphasis and should be made interactively. Otherwise, either the long-range objectives "set" at the top cannot be attained because of limitations at the bottom, or the means and goals selected at the bottom fall short of what can and ought to be attained.

Thus, integrated planning is planning that takes place simultaneously and interdependently at all levels of a system. It involves explicit consideration of all four types of decisions at each level. Participants at every level have a right to be heard on every type of decision.

Continuity. In an age suffering from "Future Shock" it is widely known that neither purposeful systems nor their environments stand still while planners try to address them. They change continuously, often in unanticipated ways. Furthermore, their rates of change appear to be increasing. Therefore, every plan, if left alone, deteriorates when such changes occur. If, however, planners learn from these changes and use what they learn to revise and update their plans, then they can at least maintain and may improve the effectiveness of their plans. For this reason, planning should be a continuous process, not a "sometime thing;" and plans should be treated as nothing more than progress reports.

The Content and Phases of Planning

The content of planning is a set of interacting problems: a system of problems. This is yet another way of saying that one cannot decompose planning into a set of independent problem-solving exercises, and that the quality of a plan depends more on how the solutions to the constituent problems interact than on how good they are when considered independently of each other. But no one can deal simultaneously with a large number of problems. Therefore, the planning process must be divided into phases, all of which should be treated interactively.

Because of the complexity of the planning process it can be divided in many different ways. Each is a different way of slicing through the same process but each slice gives a different view of the whole. We have identified five phases of planning which derive from our experience in planning. They are ones we have found useful. They are:

(1) *Ends Planning.* This phase of planning is concerned with identifying, defining, and producing measures of the degree to which ends (desired outcomes) are attained. Ends are usefully classified into goals, objectives, and ideals. *Goals* are ends whose attainment is planned within the period which the planning covers. *Objectives* are ends whose attain-

ment is not expected until after the period which the planning covers, but progress toward which is expected within that period. Goals, therefore, can be considered to be means to objectives. Objectives can, in turn, be considered to be means to *ideals* which are ends that cannot be attained but progress toward which is unlimited. For example, a student entering high school may plan to graduate within three years, his goal. His objective may be to enter and graduate from college. His ideal may be a fully educated and rounded life.

As previously noted, normative planning is concerned with ideals, strategic planning with objectives, tactical planning with goals, and operational planning with means. These are not separable, each is implicit in the others. Unfortunately, however, they are seldom carried out explicitly and interactively, as they should be.

Ends planning should involve the design of a desirable future. It is in this context that idealized design should take place.

(2) **Means Planning.** This phase of planning is concerned with what is to be done by the system planned for, with selection of means by which ends are to be pursued. Means include actions (one-time choices), practices (repeated actions), programs (combinations of actions directed at a particular goal or goals), processes (sequences of actions similarly directed), and policies (rules for selecting any of the preceding types of means).

In our view the key to effective planning is not so much the selection of the best from an available set of means as it is the extension of the set to include one or more means that are significantly better than any that were initially available. It is in this sense that planning includes inventing ways of creating a desirable future.

(3) **Resource Planning.** This phase of planning is concerned with determining (a) what resources will be required to pursue the specified ends by the means selected, (b) how they are to be acquired or generated, and (c) how they are to be allocated: who is to use them and for what. Resources include personnel; plant, equipment, and facilities; materials and energy; money; and information, knowledge, and understanding.

(4) **Organizational Planning.** Such planning is concerned with designing or redesigning the organization that is to carry the plan so that it can be done effectively and efficiently. This includes design of the subsystem for managing the organization planned for.

(5) **Implementation and Control.** This phase of planning is concerned with (a) scheduling the steps to be taken in implementing the plan, (b) determining who is to take each step, and (c) determining how each feature of the plan, and the plan as a whole, are to be evaluated and improved over time.

One or more of these aspects of planning are usually omitted. In some cases different aspects are carried out independently of each other. Both of these sins should be shunned.

Idealization in Planning

Idealized design is the core of normative planning. Only by means of such design can assurance be obtained that the pursuit of short-range goals and middle-range objectives will yield long-range progress. It reduces the chance that a system will be tempted by immediate gains that can result in ultimate losses.

If no more than idealization were to be done, nothing would be accomplished. The idealization would remain a dream, however consciously carried out. Therefore, it is intended to initiate but not to consummate a participative, coordinated, integrated, and continuous planning process. Those who engage in idealization usually do so with the hope of producing a design that will act as a mobilizing idea. The power of such an idea was poetically described by the Spanish philosopher Ortega y Gasset (1966:1):

... man has been able to grow enthusiastic over his vision of ... unconvincing enterprises. He has put himself to work for the sake of an idea, seeking by magnificent exertions to arrive at the incredible. And in the end, he has arrived there. Beyond all doubt it is one of the vital sources of man's power, to be thus able to kindle enthusiasm from a mere glimmer of something improbable, difficult, remote.

THE STUDY DESIGN

Our work to date has consisted of two overlapping phases.

(1) *Investigation of the Current System.*

This effort began with collection and synthesis of available information about the current SCATT System. We have tried to organize this information into an informative description of the system. Our purposes were (a) to synthesize the work of those who have already described aspects of the system in detail, (b) to reveal where there is a lack of adequate information, and (c) to identify deficiencies in the current system.

(2) *Idealized Redesign of the System.*

Enough has been said about the what and why of idealized design in preceding sections of this chapter. What remains to be described is the how and who of the design process.

The first version of the idealized design was produced by the project team. It was disseminated as widely as possible for criticism and suggestions. There was no lack of either. The design was then revised to incorporate as many of the criticisms and suggestions as the project team and its advisors believed were well-founded. This included most of them. In this way a succession of versions was prepared.

Reviews of our work have been both formal and informal. On the formal side, use has been made of three advisory groups set up for this purpose by the Office of Science Information Service (OSIS).

First, there was an industrially oriented group chaired by Walter Carlson of IBM. Its members were:

David Allison, Manager of Technical Communications, Xerox Corporation
Joseph Becker, President, Becker and Hayes, Inc.
Kenneth W. Lowry, Director of Libraries and Information Systems, Bell Telephone Labs
Jerome D. Luntz, Vice President Planning and Development, McGraw-Hill Publishing Company
Gustavus Simpson, Battelle Memorial Institute
Davis Staiger, Director of Publications, American Institute of Aeronautics and Astronautics
Martha Williams, Professor, Information Retrieval Research Laboratory, University of Illinois

Second, there was an academically and governmentally oriented group chaired by Professor Jordan Baruch of the Tuck School, Dartmouth College. Its members were:

Martin Cummings, National Library of Medicine
Alan J. Goldman, National Bureau of Standards
Alexander G. Hoshovsky, R&D Information Officer, Office of R&D Plans and Resources, Department of Transportation
J. C. R. Licklider, Advanced Project Research Agency, Department of Defense
Ben-Ami Lipetz, Citizens Urban Information Center
Fred K. Menasse, Drexel University
Arthur Miller, Harvard Law School
Y. S. Touloukian, Purdue University
F. Karl Willenbrock, National Bureau of Standards

Finally, an advisory group made up of faculty members of the University of Pennsylvania who were not otherwise involved with the project set up. Its members are:

Donald C. Carroll, Dean of the Wharton School and Professor of Management
Ronald Frank, Vice Dean of the Wharton School and Professor of Marketing
Thomas Hughes, Professor of the History and Sociology of Science
Klaus Krippendorff, Associate Professor of Communications
Hasan Ozbekhan, Professor of Social Systems Sciences
Eric L. Trist, Professor of Social Systems Sciences

In addition, many others played a role in the design presented here. We are grateful to all of them.

CONCLUDING REMARK

The idealized design of the National SCATT System is presented in Chapter 2. In Chapters 3 to 5 we elaborate upon the operation of its coding and classification system, the feasibility and availability of the technology required, and its economics. Chapter 6 outlines the way in which this National SCATT System would fit together with those of other countries.

It is consistent with the philosophy of planning set out here that the last chapter (7) provides a description of the non-system of today from which we must move forward. This summary is included primarily for purposes of comparison, for its contents will be well known to many of our readers. Those, however, who are relatively new to the field may prefer to read it next, before turning to Chapter 2.

2

DESIGN
An Overview of the Idealized SCATT System

INTRODUCTION

An idealized design is based on a set of values placed on desired ends and a set of assumptions about the efficiencies of means to these ends. Our experience shows that these values and assumptions, these attitudes and beliefs, are not usually formulated explicitly at the beginning of the design process. They are normally extracted by analysis of the design once it has been tentatively completed. Then they can be examined and evaluated in a relevant context. Such evaluation usually feeds back into the continuing design process and can be repeated at the end of each cycle. We followed such a procedure in developing the design presented here.

This ordering of events enables participants in the idealized design process to express their values and assumptions operationally and in a synthetic mode, as parts of a whole, rather than in an analytic mode, as independent components of the design. This makes it possible for the values and assumptions to interpenetrate, illuminate, and enhance each other.

OUR ASSUMPTIONS

We introduce our idealized design by presenting a list of values and assumptions that underlie it. We do so because it may facilitate understanding and (hopefully) appreciation of the design. Some of the values are broad and societal, some are narrow and more directly applicable to the SCATT System.

The values and assumptions involved in our design are as follows:

(1) That there is a societal need for better results from science and technology and that this need is serviced by increased efficiency of scientific communication and technological transfer, which reduces unproductive duplications of effort and leads to creativity and innovation.

(2) That the good or evil done by man with science and technology depends on the extent to which applications of these are controlled relative to man's needs and desires. Such control, in turn, depends on the extent to which the content and consequences of the output of science and technology are known and understood by scientists, technologists, and the public at large.

(3) That the effective and efficient production and use of science and technology can be facilitated by communication within and between the scientific and technological communities and the larger communities that contain them.

(4) That a SCATT System which produces the kind of communication that facilitates effective and efficient production, distribution, and use of science and technology must be oriented primarily toward its users. It should also effectively serve other participants and stackholders in the System; for example, producers, publishers, and disseminators.

(5) That the System should be flexible and adaptive to changes in (1) the needs of users, other participants, and stakeholders, (2) the social environment, and (3) the relevant science and technology.

(6) That the System can best be made responsive to the needs of those it serves both directly and indirectly by (a) enabling them through direct or representative participation to effect its continuous redesign and management, (b) requiring the system to support itself through charges for its services, and (c) permitting competitive services to come into existence. In other words, we assume that participative management and the market mechanism are effective ways of providing the SCATT System with continuous evaluation of its performance by its participants.

(7) That the time of working scientists and technologists is a valuable social resource; hence, that the amount of it required to produce, distribute, gain access to, and use information should be minimized. That this can be accomplished in part by reducing the amount of irrelevant, redundant, and useless information they receive and by providing as much relevant, new (to them), and useful information as they want, but no more than this.

(8) That the primary responsibility for determining the relevance, redundancy, and usefulness of information should rest with its

users, and that the System should enable them to make such judgments accurately, reliably, and with minimal expenditures of money and time. Put another way: we believe that the System should not make choices for its users, but that it should make available to them choices and ways of making them that would not otherwise be available; it should expand, not restrict, their freedom of choice and their capability of choosing well.

(9)   That the SCATT System should be easy to use in a wide variety of self-determined uses; and that it should facilitate its users' learning how to use its services effectively and efficiently.

(10) That the services provided by the SCATT System should be available to all who desire to use them.

(11) That the SCATT System should facilitate informal as well as formal communication; in particular, that it should encourage the formation and operation of informal networks or "invisible colleges" and provide unobtrusive access to them.

(12) That the SCATT System should protect the privacy of individuals and enable them to minimize the amount of unsolicited information they receive.

(13) That scientific and technological information, knowledge, and understanding should be as publicly available as possible, minimally restricted only for reasons of personal privacy, national security, and preservation of proprietary rights.

In addition there were two constraints which we accepted:

(1) *Because of the mission of the sponsor, the Office of Scientific Information Service (OSIS) of the National Science Foundation (NSF), this is a design of a National SCATT System.*

But there is nothing basically national about the needs this design addresses. We have tried to make it sufficiently flexible to admit interaction with other national systems in an international network of which we provide a sketch in Chapter 6. Indeed, one of the objectives of this study is to stimulate similar idealized design efforts in countries other than the United States. Some steps in this direction have been taken.

(2) *Our discussion of the design is limited to science and technology because of the sponsor's interest and concern, but we believe the design is applicable to the arts, humanities, business, government, and fields other than science and technology.*

There are, however, certain aspects of scientific and technological communication with which we do not deal. Our exclusions are best understood by reference to Table 2-1. The rows and columns of Table 2-1

Table 2-1.
Categories of Scientific and Technological Communication°

Destinations

Sources	*1. Science (Research)*	*2. Technology (Development and Design)*	*3. Practice*	*4. Ultimate Use*
1. Science (Research)	11	12	13	14
2. Technology (Development and Design)	21	22	23	24
3. Practice	31	32	33	34
4. Ultimate Use	41	42	43	44

* This classification was suggested to us by Walter Carlson (IBM). A similar classification of information users appears in Organization for Economic Cooperation and Development (1971): (1) the scientific specialists (the *researcher*), (2) the industrial engineer (the *applier*), (3) the planner, the policy maker, the decision maker, and the manager (the *innovators* and *guiders*), and (4) the *public* (the consumer, beneficiary, and victim).

represent functions, not persons. An individual may be involved in all four functions, and usually is at various times. A research physicist, for example, may be an ultimate user of computer technology and practices as well as of the scientific research on which they are based. Nevertheless, most individuals are predominantly associated with one of these functions. Furthermore, those in each function direct most of their communications to others who are predominantly in the same function. For example, operators of computing centers or medical doctors direct most of their communications to others in their same function. Therefore, communications within cells 11, 22, 33, and 44 are the most highly developed. Communication to an adjoining cell on the right is the next most developed; for example, from scientists to technologists, from technologists to practitioners, and practitioners to users. There is generally less communication to adjoining cells on the left and still less to several cells away in either direction.

The SCATT System presented here focuses on the upper-left quadrant (11, 12, 21, and 22), and particularly on 11 (from science to science) and 22 (from technology to technology), though less on the latter than the former. To a still lesser extent it deals with the remainder of rows 1 and 2. It does not deal explicitly with rows 3 and 4, communications originating in practice or use but the System is capable of receiving and distributing such communications.

CONCEPTS AND CATEGORIES

Every system design (and description) is based not only on a set of assumptions and values, but also on a set of concepts and categories that direct it by establishing criteria for relevance and that provide a means for organizing the results. It was necessary, therefore, to begin this effort by specifying the concepts and categories that we intended to use. In doing so we relied heavily on the output of a study previously carried out under the sponsorship of OSIS (GN-389). The final report of this study was called *Choice, Communication, and Conflict* (Ackoff, 1967). The content of this report was subsequently expanded, modified, and incorporated into the book *On Purposeful Systems* (Ackoff and Emery, 1972).

Here we identify and define (with less rigor than is provided in the earlier report) those concepts on which our design and descriptive efforts depend most heavily.

(1) *Sign:* anything that is a potential producer of a response to something other than itself.
(2) *Message:* a set of one or more signs intended by its producer to produce a response either in another or himself.
(3) *Communication:* a change in one or more of the parameters of an individual's purposeful state produced by a message.

The parameters of an individual's purposeful state are as follows:

(1) *His probability of choice* of each of the available courses of action.
(2) The *efficiency* of his possible choices; that is, the probability of their producing each of the possible outcomes.
(3) The *relative value* of each possible outcome to the individual.

Therefore, there are three *modes* of communication.

(1) *Information:* a communication that produces a change in any of the receiver's probabilities of choice, informs him.
(2) *Instruction:* a communication that produces a change in the efficiencies of any of the receiver's courses of action, instructs him.
(3) *Motivation:* a communication that produces a change in any of the relative values the receiver places on possible outcomes of his choice, motivates him.

A single message may have any one, two, or all three of these effects.

Shannon and Weaver (1949) and many who folllowed them used "information" in a different sense from the one provided here. What Shannon and his predecessor Hartley (1928) referred to as the amount of information in a message is the amount of message, not of information in it. Shannon did not distinguish between information, instruction, and motivation; hence "information" has come to be used in technical discourse to cover all three types of message content. We do not depart from this now common technical usage, but we take "information" to include what we have explicitly designated as instruction and motivation.

Information, instruction, and motivation are conveyed by signs. Those signs that represent properties of objects and events are called *data*. *Useful* data—data that can affect the behavior of the receiver—are information. Data usually have to be processed to be made useful, to be converted into information. Therefore, we use "data" to designate the raw material out of which the finished product, information, is made.

Measures first developed in 1958 of the amount and value of information, instruction, and motivation are provided by Ackoff and Emery (1972). An application of some of these measures is described by Martin (1963).

We also make heavy use of the distinction between *auditory* and *visual* signs. A single message may obviously combine both—for example, one delivered over television. Written messages are only one type of visual communication. Gestures and photographs can also be used to communicate visually.

The terms *formal* and *informal* are used extensively in discussions of scientific communication and technology transfer. Taken literally, these terms refer to the imposition of form on communication, such as limits on length and rules covering format. The formality itself, however, does not seem to us to be nearly so important as two other properties that are usually associated with it: *asymmetry* and lack of *spontaneity*. Formal communication usually restricts or prohibits feedback; informal communication normally encourages it. A letter (informal) invites a reply; the reply in turn invites a reply, and so on. In this sense an exchange of letters resembles a conversation, which is the most informal type of communication. But an article or lecture (formal) tends to restrict or prohibit feedback.

Formal communication also usually involves preparation, hence it lacks spontaneity. An extemporaneous lecture, or even one delivered from notes, is less formal than one read from a prepared paper. A spontaneous letter is less formal than one that has been carefully prepared, such as a form letter.

Symmetry and spontaneity are important because of their effect on the effectiveness of communication from the receiver's point of view. Formality is usually imposed on communication by its producer or disseminator, seldom by its receiver.

Therefore, we use "formal" to refer to communications that tend toward asymmetry and preparedness and "informal" to refer to those that tend toward symmetry and spontaneity. There is no sharp line between them and communications may vary widely with respect to both "dimensions."

By combining the last two sets of categories we obtain four types of communication:

(1) *formal auditory*—for example, a prepared lecture;
(2) *informal auditory*—for example, a conversation;
(3) *formal visual*—for example, reading a published article; and
(4) *informal visual*—for example, an exchange of personal letters.

In communication systems involving recorded transmittable messages, messages may be of three types:

(1) *primary:* messages believed by their producers to convey information not previously communicated—for example, papers and lectures reporting new findings;
(2) *secondary:* messages about primary messages that affect awareness of their existence and availability—for example, bibliographies and indexes; and
(3) *tertiary:* messages about the content of other messages—for example, abstracts, reviews, digests, and state-of-the-art papers.

A single communication may contain all three types of messages. Any one of the three types may be auditory or visual, formal or informal.

Finally, we divide communication involving recorded messages into four phases.

(1) *Production:* all processes by which a message is produced in transmittable form and entered into a distribution system—for example, the writing and publication of an article and the preparation and delivery of a lecture. In technology, the product often is not a message but an object (a good) or a process. However, messages are often produced that describe them (e.g., patents) or how to get access to them (e.g., catalogs).
(2) *Dissemination:* all processes by which a message is made available to potential users and by which they are made aware of its availability—for example, entry of a document into a library, cataloging and indexing it, or notifying potential users of its location.
(3) *Acquisition:* all processes by which an individual gains access to messages—for example, by withdrawing a book from a library, subscribing to a journal, or sending for a reprint.
(4) *Use:* the production of an effect on an individual by a message, that is, receipt of its content.

It should be emphasized that these four phases are functions that may be combined in various ways in a single activity or in a single individual. No individual ever performs only one of these functions; every producer is also a user and every user produces something. We must discuss these functions separately because of the limitations of the written word, but it should constantly be borne in mind that the functions are highly interactive and that scientific communication and technology transfer are not "linear processes" with clear-cut beginnings and ends. They are inter-penetrating and tend toward continuity.

Now we turn to the idealized design itself.

ESSENTIAL FEATURES OF THE IDEALIZED DESIGN

The presentation of the essential features of our idealized design is organized into twelve sections:

1. Production and entry into the System of documents and information
2. Meetings and conferences
3. Acquisition of information and documents from the System
4. Feedback to and from users
5. Storage in the System
6. Fellows, annual reviews, and consultation
7. Economics of the System
8. Research, development, and education
9. Special services
10. International connections
11. Organization of the System
12. The role of libraries

The diagram shown in Figure 2-1 represents the essential flows in our idealized SCATT System. Each element of the diagram is discussed in detail below.

A source may communicate findings, thoughts, and questions either orally or in writing or both, within or outside the System. Private communications, either written or oral, that are initiated by the source are not discussed further here except insofar as they employ the facilities of the SCATT Communication Network. It is assumed that they would continue much as now except for use of new communication technology such as the picture phone and facsimile transmission. On the other hand, private communications from the user to other participants in the system are taken up in detail.

We number each paragraph (¶) of the design features for easy reference. Our comments on these features and their presentation are contained in italicized paragraphs to differentiate them from description of the design itself.

Production and Entry into the System of Documents and Information

1. An author or authors might submit a manuscript either to a publisher or directly to the National SCATT Center. In either case the manuscript

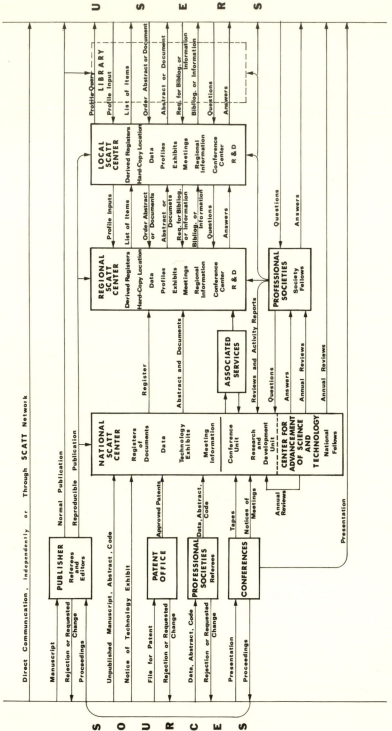

Figure 2-1. Schematic Diagram of Main Features of an Idealized SCATT System.

would be required to be either in machine-readable form or in a format that facilitates conversion to such a form. The author(s) would also be required to provide an index and an abstract of from 100 to 250 words and to indicate by use of appropriate coding and keywords (discussed in Chapter 3) the nature of the document's content.[1] Automated indexing, abstracting, and coding facilities would be available through Local SCATT Centers but their use would be optional. Standard forms on which to provide this information would be available at all SCATT Centers and affiliated institutions and libraries. In the case of documents containing separately authored or otherwise independent parts (e.g., collections of papers), coding, key words, and abstracts would be required for each part. If an article is submitted without appropriate coding, classification, and so on, the System would provide it on request but at a cost to the one who submits it.

Note that an author could enter his manuscript into the system before publication or without it. This would not only provide a significant alternative to publication but it would also apply pressure on publishers to accelerate publication.

2. Patent applications would be submitted to the Patent Office in a specified form accompanied by appropriate coding, keywords, and a 100-250 word abstract, also on a standard form.

3. Those wishing to enter a manuscript, an article, or a book into the National SCATT System would have to make a redundancy check with the assistance of a Local SCATT Center or a library authorized to perform this function. Such a check would involve having the classification of a manuscript verified, then extracting abstracts of all closely related documents.[2] The author would compare his submission with these and prepare a brief statement to be added at the end of his abstract identifying what is new in his article or book. With his manuscript he would submit a copy of the list of documents he has checked and a certification of his coding by a Local SCATT Center or a library authorized to provide such certification. Each article or book would be required to include a list of articles and books with similar content that have appeared in the last five years. If there are more than fifty (or some other more suitable number of such) documents, only the fifty most recent ones need be listed. The best number to use should be settled by experimentation.

A special service would be available to enable an author to carry out a redundancy check prior to preparing a document.

---

[1] We often use the term *document* in place of the more cumbersome term *primary message*. The term *document* should therefore be understood broadly to include data, patents, tapes, and so on.
[2] The method of selection of closely related documents is discussed in Chapter 3.

4. Every citation of a document would include its SCATT identification number (discussed in Chapter 3).

The features included in ¶ 3 and 4 are intended to facilitate "follow up" by a reader of a document, including his acquisition of documents referred to.

5. Every periodical would be divided into three clearly designated parts: (a) invited papers, (b) uninvited papers that have been refereed and accepted for publication, and (c) uninvited, unrefereed, or rejected papers.

> **5.1.** There would be no fee associated with publication of invited papers.
> **5.2.** Publication of uninvited but refereed and edited papers would involve a charge to cover part, but not all, of the costs of their processing. A paper for submission of which a fee is paid would have to be rejected within a specified period of time (for example, 3 months), or else it would be accepted automatically for publication. If a manuscript were rejected by one publisher and were sent by the author to another, an additional fee would have to be paid. If revisions were requested by the publisher and a subsequently modified manuscript were resubmitted, a reduced reprocessing fee would be required. A rejected paper could be resubmitted to the journal that rejected it as a "not to be refereed" paper.
> **5.3.** Publication of uninvited, unrefereed, or rejected papers would involve a charge to cover the total processing cost. If more such papers were available to a journal than it could publish, a random selection would be made. A journal could not reject an uninvited paper unless at least one-third (or some other suitable portion) of its pages were devoted to papers that had been rejected or unrefereed.

The reasons for the charges is to discourage casual publication, to give an author pause before rushing into print. The reason for requiring publication of unrefereed and of refereed-but-rejected papers is to provide journal editors with a way of evaluating their refereeing processes. The SCATT System would be able to provide them with statistics on users' evaluations of articles in all three categories. The procedures specified above would also provide any author who is convinced of the value of his article despite its rejection by referees with a chance for its publication. Because rejected articles might exceed the space available for their publication we provide yet another way for the "convinced" author to enter his paper into the SCATT System (¶ 13).

6. Publishers would be responsible for carefully checking the completeness and correctness of coding, key words, and abstracts. Every published article and book would be preceded by the author's abstract and another abstract arranged for by the publisher and prepared by someone who has not seen the author or machine-prepared abstract. The author of the second abstract would be identified in the publication.

Comparison of these abstracts would provide a potential user with a better basis for deciding whether to read a document than he currently has. It would also discourage authors from using abstracts as "commercials."

7. The Patent Office would carefully check the coding, key words, and abstracts of the patents it accepts.

8. One who had accumulated numerical data that might be useful to others could submit such disembodied data to the National SCATT Center in a machine-readable form specified by that Center. Such data would have to be accompanied by a verbal description of its character and how and when it had been obtained. It would also be coded and described by key words. Before submission they would have to be refereed by an appropriate professional society or agency thereof. The data would be entered on a *National Data Register* without charge unless it were copyrighted, then the cost of entry would have to be paid by the author or the professional society involved.

9. All published documents, tapes, and accepted patents would be required to be sent to the National SCATT Center where their identification and coded classification and key words would be entered on the *National Document Register*. Abstracts and, in the case of books, Tables of Contents would be entered into the *Secondary Information File*. The cost of such entries would be borne by the publisher or author.

10. If a document already in the System in manuscript form were published (in a journal, for example) then the manuscript would be removed from the System. The coding and identification of that manuscript would, however, remain in the National Document Register, and, if requested, automatically linked to the published material. Statistics on use of that manuscript would be transferred to the published version. Similarly, a new revised edition of a document would replace the old one; the old version, however, would not be removed from the system, but would be dropped to a lower storage level.

The publisher would thus have a major incentive to shorten the delay between the preparation of the manuscript and its publications. If that delay were too great, most of the potential readers of a document would have read it in manuscript rather than in published form.

11. A good deal of primary information about technology is transmitted by observing the technology itself—for example, at exhibitions or the location at which it is installed. Information about the possibility of such observation would be submitted to the National, Regional, and Local SCATT Centers and be retained in a *Technology Exhibit Register*. The contents of this register would be continually updated.

For example, one might want to find out where he could see a particular type of oxygen furnace or computer installation. The register would identify the places where he could do so and whom to contact in order to make arrangements to do so.

12. Authors would send corrections of errors in documents to the National SCATT Center where the corrections would be made in the stored original. The entry in the National Register would be modified so as to include the existence and date of the correction. All other centers storing the entry would receive the modification. The cost of such corrections would be borne by the author or his publisher. For this purpose documents would be partitioned into numbered paragraphs (as this chapter is) in order to make identification of its parts precise and independent of format. For example, 12-5-7 would represent the seventh word in the fifth sentence of the twelfth paragraph. Indexes of books and articles would also use this location-identification scheme.

13. Every journal would have a section devoted to (a) letters and commentaries from readers, (b) statements of problems deserving attention, and (c) presentation of "half-baked" ideas and conjectures. Each such entry would be appropriately coded and accompanied by key words. In particular there would be information on whether it had been invited, uninvited but refereed, or uninvited and unrefereed.

14. In order to encourage and accelerate release of "classified" scientific and technological information into the public domain, two measures would be taken:

14.1. The National Academy of Science and the National Academy of Engineering would establish review boards to determine whether governmentally classified documents should or should not remain classified. If not, and they were judged worthy of publication, they would be released to the SCATT System and registered at government expense. Such reviews would take place three years after preparation of the document and be repeated every three years thereafter for documents left in the classified category. Members of the review boards would be appropriately "cleared" but would not be in the employ of the government. In the case of a disagreement over the classification, the burden of proof would rest with the government agency.

14.2. Private organizations would receive tax credit for documents describing research or development they have carried out that appear either as invited or refereed articles or as books.

15. Each professional society would publish a periodical that appears no less frequently than once per month that contains descriptions in 500 words or less of recent results of research or development, or their applications. The source of each description and its address would be provided. Small societies that could not justify or sustain such an effort on their own would collaborate with other societies of related interests in producing a shared periodical of this type.

Meetings and Conferences

16. A source might deliver a paper at a conference, meeting, symposium, or other form of face-to-face interaction. (Such presentations would also occur through two-way audiovisual media.) Proceedings of such formal events could be recorded on reproducible audio or audio-visual tape, copies of which could be made available rapidly through the SCATT System.

17. Each tape entered in the System would require a written abstract and coded classification. If presentations thus recorded were subsequently published in documentary form, the corresponding tapes would be withdrawn from the system's "active" inventory but, nevertheless, would remain available.

This would make it possible for historically significant presentations to be heard and seen long after they had been made.

18. Each Regional SCATT Center would have conference facilities associated with it. These would be available with charge to any scientific or technological organization. They would be equipped so as to enable effective experimentation with the form and content of meetings conducted in them. The research and development (R&D) staff of the Regional SCATT Center would work with and assist those conducting meetings in these facilities in designing sessions that would effectively accomplish their meetings' objectives. Each meeting held in SCATT facilities would be systematically evaluated.

19. Each Local SCATT Center would have facilities available for smaller meetings. These would be similarly equipped and augmented by Local SCATT Center R&D personnel.

Conferences have been a major target of complaints for many years. It is unlikely that there is any such thing as a best way of conducting them. Different formats best serve different purposes. Therefore, in this design we cannot specify an ideal conference, but we require systematic experimentation and evaluation of different types of conference, so that conference planners can learn quickly and effectively from their own and others' experience.

20. Coordination of experiments on conferences and dissemination of information about them, including evaluations, would rest with the R&D unit of the National SCATT Center, which would have a conference department. Its services would be available on a consulting basis.

21. Each SCATT Center would maintain for the geographical area it serves a *Meeting-Information Register* in which information about future meetings of all kinds would be recorded. The subjects of the meetings would be appropriately coded. The record would contain the subject, sponsor(s), time and location, fees, and an indication of whether a tentative or final program was available. Programs would be submitted in machine-readable form. The registers would regularly be purged of information on meetings that have already been held.

22. Senior scientists and technologists and young associates or students whom they take with them to conferences would be subject to lower registration fees than otherwise.

Such reduced "group registration fees" are intended to facilitate absorption of young scientists and technologists into their mentor's invisible colleges (informal networks) and formation of their own.

23. The System would store, update, and disseminate descriptions of available science- and technology-based college curricula and learning opportunities offered by public and private educational institutions.

Acquisition of Information and Documents from the System

Before considering how a user could obtain information and documents he wanted, we include a feature to reduce his receipt of information and documents that he did not want.

24. In order to decrease the volume of unsolicited information that scientists and technologists receive, there would be a higher distribution (e.g., postal) rate for any unsolicited document other than non-form letters.

This reverses current procedures in which most unsolicited documents are posted at a lower rate than private correspondence. If a receiver agrees to be placed on a particular organization's mailing list for all or a designated part of its output, that organization could use the lower rate. But if it passed on its list to another organization, that organization would have to use the higher rate. Any receiver of an unsolicited communication sent to him at an improper rate could report it to the Post Office, in the case of mail, or to the SCATT System, in the case of messages transmitted over its network. Such infractions, once confirmed, would result in a fine imposed on the sender. If the objection were overruled, the one making the report would be fined.

The objective here is clear: to reduce the amount of "junk mail" and unsolicited documents received. It would be possible, for example, for a publisher to write (not mass-produce) letters to individuals asking them if they wanted to be on a specified mailing list. Mailings from professional societies of which one is a member would be considered to be solicited.

Now we consider acquisition of desired information and documents. The core of the proposed distribution system is a profile analysis of individual and institutional users who desire the services described below. This system would work as follows:

25. An individual or institution that wanted to use the profile-based service would receive a complete list of categories and key words used in the National Document Register. He would construct a profile (as described in Chapter 3) using those categories and key words of interest to him. Once he had prepared a tentative profile, his Local SCATT Center or an authorized library would provide him with a sample (of a size to be determined) of recent documents that his tentative profile would have extracted from the System. After examining them he could revise his profile appropriately. This dialogue with the System could be continued until the applicant is satisfied with its yield.

26. The user would also receive additional copies of the list of categories and key words. He could send these to associates of his choosing and ask them to select categories and key words they believed characterized his interests. He could then use their selections in preparing his profile.

This procedure would take account of the fact that others often know better what our interests are than we do, but that we should be the ones to decide whether or not they do. The user would retain final control over his profile.

27. A third option open to the subscriber would be to have his profile prepared by a Local SCATT Center that uses his responses to documents randomly selected from broad categories in which he indicated interest.

28. A user, particularly an institutional user, could enter several profiles to reflect several different areas of interest. Charges would be related to the number of profiles submitted.

29. The profile(s) of the subscriber would be entered on the *Profile Register* of his Local SCATT Center. His profile(s) would not be made available to anyone without his permission. Other information about a subscriber would also be released only with his explicit permission.

30. If a user so desired, he would periodically receive from his Local SCATT Center a list of all documents, tapes, patents, meetings, or technology exhibits that had been registered since he received the last list whose classification intersected with his profile. The user could select the form in which the list would be provided; for example, printed, microfiche, or machine-readable file. Each form of output would be appropriately priced.

The fact that the list provided might be either too long or too short would induce user-initiated modification of his profile or of the frequency of obtaining the list.

In addition, the user could impose upper or lower limits on the number of citations he receives.

31. A user could include with his profile(s) specification of authors, publishers, publications, or any other element of document production that he wanted to have either automatically included in or excluded from the list he received.

For example, this would enable a user to exclude from the list he receives any articles that appear in journals to which he subscribes as well as information from sources he considers to be of low quality. He could also specify an author all of whose works he wanted to know about. A group of users might share a common set of such specifications.

32. An individual could use the profile-based retrieval services on both a regular subscription and a "one shot" basis. For example, he could have a regular subscription for documents on "Management Science," but for a particular research task he could ask for a survey of available literature on "Library Automation."

33. A group of individuals could develop and use a common profile. They could use any means they desired in order to formulate such a profile. Several combinatorial and statistical algorithms would be available to assist them in developing it. (Some of the techniques discussed in Chapter 3 could be used.) The choice of a particular strategy would depend on economic considerations. Note that an organization could have a single profile embracing a large class of documents. It could receive the output in machine-readable form and break it down for its individual users.

34. A reduction of the cost of retrieval would be obtainable by clustering profiles. Similar individual profiles could be clustered by means of algorithms discussed in Chapter 3. A "mean" profile of each cluster could be determined and used for retrieval purposes.

Such a mode of operating would reduce selectivity of the search, but because of its reduced cost, many regular users might want such clustering. If sufficient time were available, even some ad hoc searches could be clustered.

35. Charges for profile-based retrieval services would reflect the number of profiles involved, the frequency of listings provided, and the number and kind of specifications for automatic inclusion or exclusion.

36. Profiles of Local SCATT Centers would be maintained by Regional SCATT Centers much as the user's profile would be maintained by his Local Center. Its content would be the responsibility of Local Centers but Regional Centers would provide feedback much like that provided by Local Centers to their users.

37. An individual or institution could obtain a special listing of documents (e.g., for search purposes) at any time by request. A simple interactive language would be available for this purpose. Assistance in formulating such requests would be available at Local SCATT Centers or through a terminal connected to such centers. A charge would be made for such assistance.

38. Although the System would on request retrieve passages from primary messages (e.g., numbered paragraphs or portions of data from data tapes), it would not provide answers to questions about specific scientific facts or laws (e.g., the value of a physical constant) unless a specific source is designated.

This is intended to avoid use of the SCATT System as an authenticating or certifying agency. But such questions could and should be answered by other agencies through SCATT facilities or independently.

39. Specialized fact/knowledge retrieval systems would continue to exist and continue to be created. They could use the SCATT System to obtain input. They could also use the System for obtaining and filling "orders." To do so they would enter codes into the System that identified their services. When the relevance of such a code to a user's need is identified, one of the following two steps would be taken.

  39.1. If the user was interacting with the System in real time a special routine would be activated. It would ask the user if he wanted immediate access to the specialized question-answering service. If he did he would be switched to that service; his interaction with the SCATT System would be temporarily discontinued.

For example, a user could request a specific item of information such as the boiling point of a particular liquid or the number of people housed in mental institutions. If such information was available at a particular information service center affiliated with the SCATT System, the user would be so notified. He could then request direct access to this center through the SCATT Network and address his request to it. He would be charged for this service by the serving organization.

  39.2. If a user was not interacting with the System in real time, he would be given information in the System's report to him about relevant specialized service centers and how to communicate with them.

40. Responsibility for defining the interface between the SCATT System and a specialized information service center would belong to the center but it would be subject to the constraints imposed by SCATT's operating procedures.

41. It would be possible for a user to retrieve document indexes and to use them to retrieve relevant paragraphs rather than whole documents.

42. A specified set of documents could be combined into a book or other desired form on request. It could be made available in multiple copies.

For example, a tailor-made book could be produced for pedagogical purposes. Because documents would be in machine-readable form, production of compilations would not be difficult. All that would be required is a table of contents on which each document is appropriately identified. The copyright problems arising from such a mode of production are discussed in Chapter 5.

43. Upon request and payment of an appropriate fee the name of a user of a particular document could be entered on a special Follow-up Register that would enable him to receive automatically any corrections, reviews, commentaries or primary messages based on that document as they become available. Such a request could specify a time period at the end of which he would be removed from this register.

44. Listings of documents that have been on the National Register for a year or more would include summary information on the quality ratings given by previous users (¶50–51 discuss these ratings).

45. The profile-based list of items a user would receive regularly would indicate whether each document listed was published or not and, if published, whether it had been invited or refereed.

46. A user could browse in the SCATT System by specifying both a number of documents or pieces of secondary information about them and the area or areas from which he wanted them. These documents or pieces of information could then be selected either randomly or by any other procedure he specified.

Feedback to and from Users

47. The profile-based list of items the user would receive would also contain items (about 10 percent) that had been selected randomly from categories that did not appear in his profile but were closely related to it. The probability of including an item in the additional list would be proportional to how closely it was related to the user's profile. This could be measured by means of the "similarity index" discussed in Chapter 3.

From the user's responses to these items the system could learn whether his profile should be modified, and how. There is further discussion of this process in Chapter 3.

48. Lists of items provided to users would be organized so that there were three boxes after each item. The user would check the appropriate one. These would be:

(1) NOT RELEVANT
(2) RELEVANT BUT DON'T WANT
(3) WANT

Where the user indicates WANT he would also indicate whether he wants an abstract, table of contents (if a book), the index, the document, or some combination of these.

49. The user would return the marked list to his Local SCATT Center (in hard copy or through a terminal) and receive the material requested through the medium he specified. With these would come a list of the items for which secondary information is provided with the following boxes after each.

(1) NOT RELEVANT
(2) RELEVANT BUT NOT USEFUL
(3) RELEVANT AND USEFUL
(4) WANT DOCUMENT OR INDICATED PARTS OF IT

He would also receive a list of documents transmitted with boxes NOT RELEVANT, RELEVANT BUT NOT USEFUL, and RELEVANT AND USEFUL after each item listed. In addition there would be boxes in which he could indicate whether he considered the quality of the document to be HIGH, MEDIUM, or LOW. He would return these lists to his Local SCATT Center. The form on which he would do so would provide space for him to indicate any misclassification to which he believed he had been exposed.

50. The information obtained from returned lists would be used in two ways:

50.1. Periodically the frequency of relevance of each category and key-word would be determined for each user by his Local SCATT Center. If he failed to find more than a specified minimal percentage of items in a particular category or key-word class relevant over a specified period of time, he would be so notified and asked if he wanted to drop the category or keyword.

50.2. A similar frequency would be determined for categories and key words not in the user's profile but included at random in the listings he received. Notices of these would enable him to add categories and key words to his profile.

We believe users would be motivated to provide honest and timely feedback because doing so would facilitate improvement of their profiles.

51. A user could prepare a commentary of any document he had used and, on payment of a fee, enter it into the SCATT System in an appropriate form. Whenever a document on which any commentaries had been submitted was listed in response to a user's request, there would be an indication of the number of such commentaries that were available. A listing of the authors with the dates and lengths of their commentaries and the commentaries themselves would be available on request.

52. These commentaries and replies to them could be used as the basis for subsequent revisions of documents by their authors. Permission for such use of commentaries would have to be obtained from their authors unless a waiver accompanied their initial submission.

53. Users and Local SCATT Centers could obtain documents and tapes either from their publishers or from another center. Publishers would provide the National SCATT Centers with machine-readable or reproducible master copies of their publications. Regional and Local SCATT Centers would obtain copies in a way described in ¶ 106. When these were reproduced by a center for a user, a fee would be charged, part of which would go to the publisher and part to the center.

54. Local SCATT Centers would maintain a register of libraries in the areas they served at which hard copies of each document on the National Register could be obtained. Each library would notify its Local SCATT Center of each of its hard-copy acquisitions. Each regional center would compile such registers into one covering its region. Users of the System would have access to information in the local and regional *Hard-Copy Location Registers* on request.

55. Journals and books would, as now, be available for purchase from their publishers, but they could be ordered and billed through the SCATT System. Physical delivery of such material would in some cases be made through existing channels of distribution (bookstores, professional societies, etc.), but it is also likely that new companies would be established to handle the distribution of material ordered through the SCATT System.

56. The subscription cost of journals published by professional societies would be separated from dues so that members would be free *not* to subscribe to their societies' journals.

Separation of subscription costs of journals from membership dues would provide useful feedback from users on journal performance. Hopefully, it would reduce the number of low-quality journals and their sizes. This would not preclude different subscription rates for members and non-members.

Storage in the System

57. The information stored in the SCATT System would be organized hierarchically. The highest level would be accessible on an on-line basis. Other levels would be accessible by request.

58. Every new document would enter the highest level and remain there for at least one year. If a document had either not been called for more than some specified number of times during that year or had not received more than a specified number of commentaries—the numbers to be determined by experimentation—it would be dropped to the next lower level. This would continue until it reached the lowest level. The number of levels to be used would be determined experimentally and might increase with the number of documents stored.

59. Secondary information on all documents except those at the highest level would be retained at one level higher than that of the document and would indicate the level at which the document is stored.

60. No document submitted to the System could be physically withdrawn from it but an author or authors could have any of their documents in the System marked "withdrawn" if they so desired. This would indicate their attitude toward it. The author(s) could, of course, submit commentaries on their own documents.

Actual withdrawal would create problems involving cross-references, commentaries, and so on.

61. Perishable information such as would be contained in the Meeting and Technology-Exhibit Registers would have an expiration date provided by the submitting agent. Such information would be deleted from the register at that date.

Fellows, Annual Reviews, and Consultation

Periodically produced syntheses, analyses, and "translations" of recently produced scientific and technological information are potentially of great use. In this section we provide a design of a procedure for producing and distributing them. As will be seen, those involved in this procedure are also given responsibility for handling inquiries that cannot otherwise be handled by the System.

62. With financial assistance from the National Science Foundation and other sources, each professional society would support a suitable number of *Society Fellows* each year. Each Fellow would be provided with an assistant. Society Fellows could select the location at which they wanted to work provided they would be accessible by a real-time communication medium.

63. Inquiries could be addressed to Society Fellows through this medium and they would provide answers or suggest sources from which an answer could be obtained. To assist them, each society would maintain a register of experts whom the Society Fellows could consult or to whom they could refer the inquirer. On those inquiries they could not otherwise handle, they would consult with National Fellows (¶ 66–67).

64. Every society and association of individuals or organizations that produce or consume science or technology would maintain a *Current Activities Record,* updated annually, of who is doing what and who is prepared to conduct briefings on what subjects where. Copies of these records would be available in machine-readable form at the National SCATT Center.

65. Each Society Fellow would prepare an *Annual External Review* of fields other than his own from which his field could benefit.

66. Each year the National Science Foundation would appoint and support a number of *National Fellows* who would cover those fields in which the quantity of documents published per year significantly exceeds the number an individual in that field can cover adequately.

67. The National Science Foundation would pay the salary and additional living expenses of the expert, an assistant, and a secretary of his choosing. They would be housed at a *Center for Advancement of Science and Technology* (CAST) operated by the National Science Foundation.

68. For two or three hours of each work day National Fellows would be available to receive and reply to inquiries only from Society Fellows.

69. Users throughout the nation would be provided with a list of National and Society Fellows currently available for consultation, along with instructions on how to reach them, directly or indirectly.

70. Each National Fellow would prepare an *Annual Internal Review* of new documentation in his field(s). These reviews would be made available by subscription or through SCATT Centers. Suggestions of topics to be covered would be given to him by the appropriate societies.

71. The National Science Foundation would also maintain a "stable" of professional writers, competent in science and technology, who would attempt to translate the annual reviews prepared by National Fellows so that they could be understood and used at each of the other three "levels" of the scientific and technological communication network.

Recall: the four levels were Science, Technology, Practice, and Ultimate Use (see Table 2-1).

For example, the Annual Internal Review *prepared, say, in high-energy physics would be edited so as to be useful to relevant technologists, practitioners, and the general public.*

Economics of the System

The economic functioning of the SCATT System is described in detail in Chapter 5. Only its key features are summarized in the following paragraphs.

72. All the SCATT System's information services would be charged to whoever used them.

The free-market mechanism would be used in order to facilitate individual and collective evaluation of services by users. Effective cost-benefit analyses of such services cannot be carried out by centralized evaluation. The "free market" would also encourage competition, particularly where demand is not well served by the SCATT System or where its service is too expensive.

73. Once set up, the SCATT System would be required to be financially self-supporting. It could not accept operating subsidies from another organization or individuals. External financial support of its operations could be provided only by subsidies made available to its users.

This is intended to assure the user-orientation of the System. If, for example, the National Science Foundation wanted to support a new service offered by the SCATT System it would be able to do so, but only by making available subsidies for the use of this service.

74. Charges for services would take into account their users' ability to pay and their needs for particular services.

Two forms of user subsidy would be available. First, outside institutions (e.g., universities or companies) would be able to subsidize particular groups of users by meeting all or part of the costs they incur in using the SCATT System or competitive services. Second, the System itself would be able to subsidize certain groups, such as students and retired persons, by offering them services at reduced rates. Such rates, however, would not be less than the incremental costs of providing the particular services.

75. Academic libraries would also be required to be self-supporting. Students and faculty members would be given an appropriate amount of credit, in voucher form, with which to pay the charges libraries would make for their services in or out of science and technology. The libraries' income would come only from redemption of vouchers and from cash payments made to them by the consumers of their services.

76. There would be profit-and-loss accounting in each SCATT Center, whether local, regional, or national. Charges for each type of service at all Centers at the same level would be the same. Data on costs and revenues, necessary for the purposes of planning and control, would be collected throughout the System.

Such data would be required for, among other things, control of costs and pricing of services.

77. Local SCATT Centers might be either publicly or privately owned. Those privately owned might be for profit.

Research, Development and Education

78. The national and all regional and local SCATT Centers would be required to use at least five percent of their expenditures in any calendar year on research and development directed to improving and extending their services. The user and his needs would be the principal focus of such research. Users would be informed of any changes that affect them or the usefulness of the system.

79. The National SCATT Center would have an *Educational Unit* that would develop programs to teach the use of the System. Courses, displays, and descriptive material on use of the System would be made available at Local SCATT Centers, libraries, and institutions requesting such service, including schools. Courses would be of three types: for new users, on new services, and refresher courses. Much of the material could be provided through computer-assisted instruction.

80. For a fee, Local SCATT Centers would monitor the use of the system by any of its users who request it, point out inefficient or ineffective usage, and suggest ways by which the users could improve their use.

81. Changes in the System would be made in such a way as to enable users to continue using the System as they had been, but they would receive the information about changes and be offered the instruction necessary to enable them to take advantage of the changes.

82. Financial rewards would be provided to those who reveal to the System ways of breaking its security. Such information would be used by the R&D Unit in the National SCATT Center to correct the deficiencies revealed.

Special Services

83. Public and institutional libraries, in addition to carrying out their current functions, would perform the following:

(1) store reproducible copies of documents and make nonreproducible copies of these documents available for users at a fee shared with the publisher, and

(2) provide staff and equipment (e.g., terminals) that would make possible use of the SCATT System.

84. Information about users and uses of the SCATT System that would be stored in the System could be helpful for purposes other than those of the System—for example, for scientific and technological surveys and market analyses of documents. Such information would be available at a fee and would be provided only in such a way as to protect completely the privacy of the users of the System.

85. Anyone wanting to make a distribution to SCATT System users having specified profile characteristics could arrange with a SCATT Center to have the mailing made for him at a fee. This would protect the privacy of the users. The material distributed would be subject to the higher distribution rate (¶ 24). Users of the System could prevent their inclusion in all such distributions or could exclude certain types of material. This service could be used effectively, for example, by authors who want to circulate prepublication copies of their manuscripts for comments or by publishers who want to announce a new publication.

86. Similarly, anyone who wanted to communicate with a specialist or specialists having specified characteristics (for participation in a conference or task force, for example) could arrange with a SCATT Center to send messages to them. The privacy of the users not wanting to receive any such messages would be protected. Those who would receive such messages would not be identified to the sender who would only be informed about those who had replied and the number to whom his message was sent.

87. The System would be able to provide universities and other employers with information about the reception of publications of those being considered for appointment or promotion. Such information would be provided only with the consent of the person being evaluated.

This service could reduce, if not eliminate, the "publish or perish" syndrome, by making quality more important than quantity.

88. Local and Regional SCATT Centers would collect and make available information relevant to the areas they serve—for example, laws, building ordinances, and codes that affect the practice of engineering in the area.

89. Facsimile transmission services (with and without color) would be available at each Local SCATT Center thereby enabling members of invisible colleges, for example, to transmit recently completed manuscripts to other members of their colleges.

90. All authors and publishers who entered documents into the SCATT System would provide addresses at which they could be reached. These would be maintained on a register and would not be released. But users of the System would be able to employ it to send messages to any author or publisher. Authors and publishers could reply by the same means. The System would protect the anonymity of any author who desired it.

91. The System would enable and encourage users and others to submit suggestions and complaints regarding its operations, anonymously if they wished.

92. Development and provision of other specialized information services (e.g., citation indexing) by public or private organizations would be encouraged by offering services at reduced rates to such organizations.

93. The SCATT Communication Network would enable any user of the System to communicate with any other user through a wide variety of media. It would also make decentralized conferences possible.

International Connections

This subject is discussed in more detail in Chapter 6.

Ideally all the scientific and technological literature of the world would be in one language. Adoption of a common language is not imminent. Barring such linguistic unity, it would be ideal to have effective machine translation of all languages into one, and back again. Such translation is not now technologically feasible. Were it not for these constraints, the design of interconnecting National SCATT Systems would be relatively easy. With these constraints it is doubtful that an effective feasible system can be designed.

94. One language would be selected to serve minimally as a common scientific and technological language throughout the world, not to replace any current language in normal use. (Hopefully it would become a universal language adopted by all fields.) Each nation would be responsible for translating its scientific and technological literature into the common language. This would be facilitated by requiring knowledge of the common language of all who pursue any degree in science or technology in any country of the world. Once all scientists and technologists were fluent in this common language, they would write in it or translate what had been written into it.

This transformation could be accomplished in a relatively short time if all, or a significant number of countries, agreed. The language selected could be either an existing one or one specially constructed. The latter would involve additional costs but might be more feasible politically.

95. The complete set of characters used in each natural language would receive a standardized code, consisting of one-to-one correspondence between those characters and combinations of symbols of the common language. Whenever a proper name is translated into the common language, the original spelling would also be indicated by use of the code. All alphabetical classifications would contain entries in both the original and common language.

96. Each nation would have a SCATT System that operates in the common language and that is compatible with all others. These Systems would be interconnected so that each could interrogate any other, obtain a copy of its document register if it so desired, and use any other as ordinary users could.

Organization of the System

97. The SCATT System would involve the following components:

(1) the National SCATT Center (NSC),
(2) Regional SCATT Centers (RSCs),
(3) Local SCATT Centers (LSCs)
(4) Congress,
(5) Using organizations and individuals, and
(6) Affiliated organizations and institutions.

The relationship between these parts of the system is shown schematically in Figure 2-2.

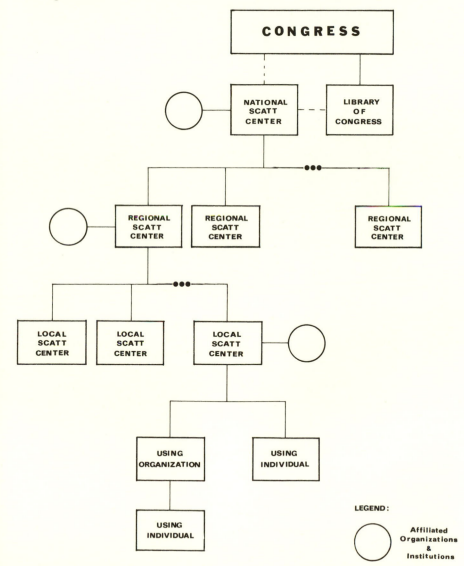

Figure 2-2.   A Schematic Diagram of the Organization of the SCATT System.

98. Initiation of the SCATT System would be authorized and funded by Congress.

Although the idealized SCATT System would be self-supporting once in operation, a considerable investment would be required to develop and implement it. This, combined with the fact that it would be a not-for-profit system, virtually requires that the initial investment be made by the federal government. Furthermore, since the NSC would require continuous interaction with the Library of Congress, it would be desirable for it to be closely associated with it.

99. The NSC and the Library of Congress would be associated but separate institutions.

Merger of the NSC and the Library of Congress was considered. The fact that such a merger might not be politically feasible is not a relevant consideration in idealized design. But there are other relevant reasons for keeping them separate, even in the ideal. Many of the functions of the Library of Congress, if not most, are not related directly to those of the NSC, and conversely. Neither is a natural part of the other; they intersect but neither contains the other. Furthermore, the skills required to operate the NSC are generally quite different from those involved in running the Library of Congress.

100. The NSC would make reproducible copies of every document in the Library of Congress that falls into the domains of science and technology. It would also carry out the necessary coding, classification, abstracting, and so on required to enter these documents in the National Register.

This would be a very large task requiring many years. Much of the investment provided by Congress would be required for this purpose.

101. A reproducible copy of every document sent to the Library of Congress for copyrighting would be submitted to the NSC. The required form, coding, abstract, and so on would be checked by the NSC. The Library of Congress would issue a copyright only if the submission requirements of the NSC had been satisfied.

102. Initially there would be about 10–15 RSCs and 100–150 LSCs. Centers could be added later as required by increased usage.

There are two principal reasons for requiring more than a NSC in the SCATT System. First, if there were only one center it would be like a national telephone exchange. The volume of communications it would have to handle would in all likelihood prohibit efficient operations. Second, even if the volume could be handled, doing so would not be as inexpensive as it would be otherwise. We have not performed the eco-nomic evaluations required to determine the optimal size of the com-munication components of the SCATT System. This should be done but it cannot be done with any precision until at least a tentative physical design of the System is completed.

In addition to logistic and economic problems that might be created by one center, there is another reason for considering decentralizing and dispersing SCATT operations. There is a good deal of information rele-vant to science and technology that is regional or local in character—for example, society meetings, lectures, building codes, ecological and energy-related regulations, and so on. These would be more effectively collected by centers in the relevant areas than by a national center. Such informa-tion, however, would be available in our system to users outside the rele-vant area by use of the SCATT Communication Network.

We tentatively propose the following simple method for allocating the LCSs. We used two requirements for determining the location of the LCSs: (1) every metropolitan area with a population of over 250,000 should have an LSC; and (2) each state should have at least two LCSs, located in its largest metropolitan areas if their populations are greater than 225,000 or, if not, located by using criteria based on actual and potential demand. (The cutoff point of 225,000 is arbitrary. It was used because it yielded what appeared to us to be a reasonable number of LSCs.)

Using these requirements we formulated a list of cities in which LCSs might be located. These cities are shown on the map of the United States in Figure 2-3.

We also propose that there be thirteen SCATT Regions as shown in Figure 2-3. There are between 10 and 15 LSCs per region.

103. On each work day the RSCs would receive all additions to the National Document Register and associated secondary information (dur-ing "off hours") and thus would have a complete duplicate of this register. By analysis of the profiles of each LSC served, the RSCs would select and transmit daily the relevant portion of the information received to the LSCs.
104. The RSC would store duplicates of the profiles served by each LSC in the region. This would provide coverage in case of a "catastrophe" at an LSC.
105. The RSCs would conduct searches for the LSCs that involve cate-gories or key-words that do not appear in the LSCs' composite profiles and, therefore, in their registers. They would also provide secondary information related to items not listed in the LSCs' registers.
106. The RSCs would receive from the NSC a reproducible copy of each document received by the NSC that intersects with their profiles. They would provide similar copies to LSCs with similar intersections.
107. RSCs and LSCs might well be located on the sites of or within exist-

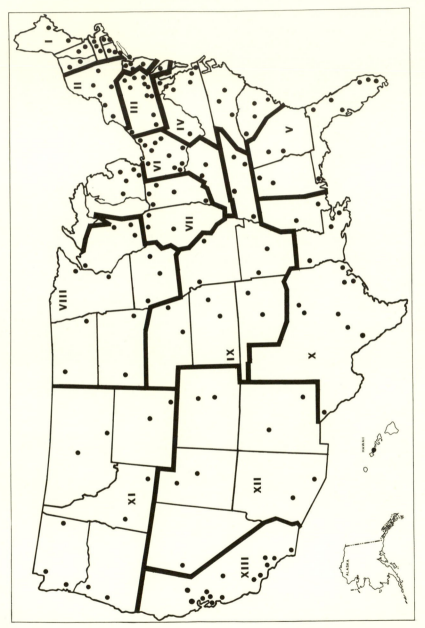

Figure 2-3. Location of LSCs and Regions.

ing institutions. For example, they might be attached to universities or research institutes.

108. The NSC would have a director and a board. The board would nominate directors who would require approval by Congress. The director's appointment would be for five years but be renewable.

109. RSCs and LSCs would also have directors and boards. The boards would nominate directors who would require approval by the next higher board. RSC and LSC directors would also have five-year appointments that would be renewable.

110. All center boards would provide representation to the stakeholders and participants. The boards would be composed as follows:

(1) The director of that center.

(2) The director of the next higher SCATT center who would serve as chairman of the board. (In the case of the NSC, its board would nominate a chairman subject to the approval of Congress.)

(3) The directors of each next lower-level SCATT Center. (In the case of the LSCs, ten representatives of the subscribers would be selected so as to provide proportional representation to each class of subscribers. The selection would be made randomly from each class. They would have staggered two-year terms.)

Affiliated organizations and institutions, whose collaboration would benefit the SCATT System, would include at the national level such organizations and institutions as the National Library of Medicine, the National Agricultural Library, the National Academy of Science, the National Research and Engineering Councils, the National Science Foundation, and the American Association for the Advancement of Science.

111. The boards would not manage centers but would have two major responsibilities: evaluation and control of the directors and establishing general policies to which the directors would be expected to adhere.

The Role of Libraries

The SCATT System designed here would clearly have major impacts on channels currently used to disseminate scientific and technological information and on the institutions that provide these channels. Libraries are currently among the most important of such institutions. Therefore, we consider here the major changes libraries would and would not experience under the SCATT System. Many of these are described elsewhere in this report, but they are gathered together here to reveal more clearly and holistically their nature.

112. Libraries would continue to provide all services they currently provide. Therefore, the traditional system would be available to those who care to use it. Libraries would continue to provide the primary interface between laymen (including students) and scientific and technological information. Their staffs would continue to provide human and informed assistance to users. Libraries would continue to offer opportunities for browsing, a place to relax or study, and hard-copy material that can be taken out on loan. Some libraries would continue to maintain essential archives, display rare books and new publications, and so on.

113. No significant change would be made in the operations of lower-level school libraries. This would also be true of most branch public libraries. Any occasional interaction they might have with the SCATT System would be conducted either through the main public library or a university library in the area.

114. Larger public libraries, special libraries, libraries in private institutions, and libraries in institutions of higher education and research would interact heavily with the SCATT System. Those that were not themselves Local SCATT Centers would in many cases install a terminal to provide access to the System. These terminals would be utilized by regular library users, by casual or occasional users who do not otherwise have access to terminals, and by users who require specialized services not available on the terminals to which they have access.

115. The SCATT System would train library staff members to assist others in the effective use of the System and to administer a SCATT facility in a library. Such staff members could conduct courses in their libraries on use of the SCATT System.

116. Public and institutional libraries, in addition to their current functions, would store reproducible copies of copyrighted documents and provide copies for users.

117. Some, if not most, local and regional SCATT Centers would be located in libraries, mainly in large public or university libraries. This would involve a significant commitment of space for which the SCATT System would pay an appropriate rent if the center is not owned by the housing institution. The space would be used for equipment, offices, conference rooms, and for providing assistance to System users and authors on formulating requests for information, conducting redundancy checks, obtaining an automatic index or abstract, programming special search procedures, and, in general, making more effective use of the System.

118. Academic institutions would need to revise their administrative procedure for funding their libraries. Their only source of income apart from fees for copying documents would be from charges for their services paid either in redeemable vouchers issued by the university or in cash. Vouchers would not have to be used at the library of the university that issued them. Therefore, such libraries might find their income decreasing if there are nearby libraries that provide more complete or better services.

This element of positive feedback is inherent in our adoption of the "free market" mechanism.

119. Libraries interacting with the SCATT System would be able to improve and expand their services. Rare, little-used, or expensive material even if remotely located could be obtained through SCATT. This would provide better and more complete services than any independent library could offer while permitting the library to concentrate its resources on the most heavily used material.

120. The SCATT System would charge libraries for the use they make of it. This cost would be passed on to the library user together with any markup the library chose to make.

No change of the magnitude involved in the design presented here can be achieved without changes in the content of the material circulating within the System. Indeed this is one of the objectives underlying the design—for example, reduction of unsolicited information. Undoubtedly the activities of the publishing industry would be affected by the users' ability to inspect document descriptions or ephemeral displays or to obtain hard-copy printouts rather than purchase or borrow a copy of a document physically produced by a publisher. A related impact would result if libraries took advantage of SCATT facilities to stock fewer hard copies of little-used or expensive material.

Thus the introduction of the SCATT System must be expected to produce changes in the type of published material available to, as well as within, libraries.

CONCLUSION

The design of the SCATT System presented in this chapter raises many questions. We address ourselves to some of them in the following three chapters, which elaborate on certain aspects of the design: the operation of the storage and retrieval subsystem; the technology incorporated in the subsystem, particularly hardware; and the financing and economy of the System.

3

OPERATIONS
An Idealized Design of Entry, Storage, and Retrieval of Information

FOREWORD

This chapter contains a detailed design of the central operations of the SCATT System. Those not interested in or wary of such detail may find it easier to proceed to Chapter 4 and return to this chapter later if they are so inclined. Chapters 4 to 7 can be read without having read this one. To those in doubt we suggest they proceed to Chapter 4.

INTRODUCTION

In Chapter 2 the general features of the idealized SCATT System's Entry-Storage-Retrieval (ESR) subsystem were described. In this chapter we present a more detailed description of it. The principal focus is on *what* this subsystem would do and on relating it to the rest of the design. We address the question of how the subsystem would operate only in order to support the feasibility of the functional design. In addition, wherever we or our advisors had an idea concerning operations that seemed worth recording, we did so.

In many aspects of this design two or more functionally equivalent options were available. Selection of the best option was not always possible without extensive experimentation and cost-benefit analyses. In such cases we have followed one of the principles of idealized design: either we incorporated the alternatives into the design or we indicate the research needed to make the choice effectively.

The Requirements of an Ideal ESR System

The purpose of the entry, storage, and retrieval subsystem is to make it possible for a user's desires for information to be matched with the documents that can best satisfy them. To make such matching possible the design of a ESR system should specify the following:

(1) how the nature and content of each document would be described,
(2) how the adequacy of these descriptions would be tested before they are entered into the System,
(3) how they would be entered into the System,
(4) how the collection of documents would be organized so as to provide easy access to them,
(5) how users would express their needs and desires so as to gain access to relevant documents,
(6) how requests and documents would be matched, and
(7) how information on the functioning of the System would be generated so as to make continuous improvement of its performance possible.

If the design of the ESR system were rigid the system could not remain effective for long because science and technology change continuously. Furthermore, the system should cater to fallible users and producers of documents, and the nature of their fallibility will change over time. Therefore, an ideal ESR system should be capable of:

(1) adapting to the changing nature and content of science and technology, including that applicable to the System itself,
(2) enabling producers and users to learn how to use it more efficiently and effectively, and
(3) adapting itself to the way it is used.

The Distinctiveness of the System

The distinctiveness of the design presented here derives from the following property: once secondary messages have been generated and stored in a suitable form, users would be able to use the data-base in any way they saw fit. The ESR system would enable its users to exercise, to the extent they so desired, all their skills and creativity in designing the services they received. However, the System would not require every user to design his own services. It would provide standard services that would satisfy many, if not most, of its users' needs.

This distinctive feature has led us to a rather different emphasis than is common in the design of ESR systems. For example, the way of describing documents becomes more important than particular retrieval methods, since it is only the former that imposes limits on what can be extracted from the System.

Coding

Before describing the design we will discuss some fundamental issues related to description of documents.

Entry of information into the System and processing requests for it requires what we call *coding*: putting the information and requests into a language that the system "understands," i.e., uses in its operations. The term *indexing* is often used in place of *coding* but we have avoided such usage because we believe it creates confusion with document indexes (which play an important role in our System) and with indexes that are surrogates for measures (for example, our Proximity and Similarity Indexes discussed later).

Preclassification versus Postclassification. Two approaches to representation of documents may be distinguished: the a priori and the a posteriori. In a priori classification (hereafter called "preclassification"), a document is placed in one or more preestablished classes. The classifier has to select from a set of specified categories those that apply to the document involved. The classes employed exist prior to the entry of the document into the System. A library's coding system is an example. The relative advantages of preclassification over "postclassification" are principally that it employs fewer categories and, therefore, is easier to learn and requires less effort to use.

Preclassification is generally based on current usage of broad categories within a field of knowledge. It relies on a traditional segmentation of the subject matters of science and technology. It respects conventional disciplines and subdisciplines as well as general trends expressed in document titles and abstracts. For these reasons it employs relatively few categories and these are generally familiar to users.

Preclassification also has some important disadvantages. It is less adaptive and becomes increasingly so with the passage of time. Because its categories tend to become less exclusive and exhaustive, their use tends to become increasingly forced and arbitrary over time.

If a new set of categories is developed to update a preclassification scheme, at least one of the following problems usually arises. First, if a new set of categories is developed independently of the older one, then the entire collection of documents must be reclassified. (This would not be practical for the SCATT System because of its size.) Second, reclassification results in inconvenience to users. To minimize such inconvenience, modification of a preclassification scheme has to assure the inclusion of the old system within the new one. This is restrictive and usually leads to an undesirably large and cumbersome system.

The a posteriori (or postclassification) approach, on the other hand, relies on descriptors of documents that are independent of any specified classification scheme. The class description of the collection of documents is derived from an analysis of the descriptors; hence it is established *after* descriptions have been prepared. As more documents are added, the classes may require modification. Therefore, a postclassification system should be capable of being modified continuously.

It will become apparent that, because the description of a document does not depend on a particular set of classes, it can be changed continuously. In this way, when the relationship of a document to others in the collection changes, this can be reflected by changing the classification scheme.[1] Many of the ways of doing so can be automated. Therefore, such changes can be less costly than similar ones in preclassification systems, which are at best very difficult to automate.

Postclassification provides more flexibility. Although it yields a larger set of categories that requires additional data processing, it makes possible, even if not practical, comparison of each document in a collection with a description contained in a request. The only limits on the specificity of a request lie in the users' ability to describe what they want.

Two general conclusions may be drawn from this discussion. First, preclassification is effective when dealing with categories that are relatively stable. It appears to be useful when dealing with the physical nature of documents (for example, "book" or "article") and broad contextual categories such as "natural science" and "life science" or, in technology, "mechanical" and "electrical" engineering because these display relative stability. It is difficult to imagine their disappearance even though new transdisciplinary categories continue to emerge.

Second, postclassification is preferable when dealing with categories that change relatively rapidly. In such cases, use of flexible categories permits restructuring the past as well as the present. It can easily adapt to work that spans established fields.

In the system we have designed, we have tried to combine characteristics of both types of classification in such a way as to preserve their respective advantages and to eliminate their disadvantages. Documents would be both preclassified and described by keywords that would make postclassification possible. This combination would facilitate a wide variety of searching procedures, ranging from ones that are rigidly preestablished to ones that can be designed by the user to serve his specific needs. The latter is possible because keywords can be used without the intervention of classes or with the use of classes of one's own design.

---

[1] The coding reflects the intrinsic properties of documents. But their properties also depend on their external relations with other documents, their extrinsic properties.

Entry, Storage, and Retrieval. Entry and retrieval clearly depend on how documents are coded. There are some who argue that coding should be completely automated. Others argue that there is no completely satisfactory way of doing so. Our design is unaffected by this argument because it places responsibility for coding messages on their producers who could select any means of coding they desired, including automated procedures if they were available.

For efficient retrieval the code incorporated into a postclassification scheme must satisfy two requirements: (a) syntactically it has to be simple enough to allow easy and inexpensive manipulation of large amounts of data, and (b) semantically it has to facilitate good descriptions of primary messages. Balancing these two requirements is not easy. We have tried to do so by use of weighted *keywords.*

As previously noted, our design provides standard search procedures for those who do not desire to create their own. Given the nature of the code used, and the current state of the art, the standard procedures necessarily combine two fundamental search strategies. The first, *"combinatorial,"* consists of using the categories of the code in order to filter out documents that are not desired. The second, *"associative,"* consists of using a statistical measure of association between a message's description and a user's profile or request to select appropriate messages from the reduced set. There are many possible measures of association. The one we use is simple and provides what is generally expected from such a measure.

Even in an ideal design the problem of organizing the storage of secondary messages is essentially a practical one. That is, given an infinitely powerful computer, there would be no need to organize secondary messages. Every document description could be matched against the user's profile and subjected to the appropriate selection criteria. Therefore, the need for a classification scheme arises from consideration of the cost of search; that is, of limiting the number of documents to be screened. There is another important reason for classification: to provide traditional (library-like) and conceptually useful means of identifying relevant groups of related items.

Finally, we provide access to secondary information about all documents within the collection at the point from which a search is carried out. This does not imply storage of such information at one point, but it does imply the ability to communicate with all the points at which such information is stored.

Now we turn to the idealized design.

ENTERING A DOCUMENT

1. Entry of a document into the System would require both a preclassification and a set of weighted keywords. The coding would be performed by the producers of documents (Chapter 2, ¶ 1 and 2).

Preclassification

It is not possible at this stage to provide a complete specification of the preclassification system. Such a system would have to be developed with the collaboration of experts in each field as well as specialists in information sciences. Therefore, we focus on the general characteristics of the System, but we are specific where we think it is justified.

2. Preclassification involves selecting from one or more sets of categories those in each set that apply to the thing being classified. (For example, one set of categories might be "type of document," including book, journal article, working paper, letter, and so on. Another set might be "mathematical techniques employed," including inventory theory, linear programming, queuing theory, and so on.) The categories in each set would be numbered using the sequence of powers of 2; that is, 1, 2, 4, 8, 16, 32, and so on. Then if more than one category were selected—and this would be likely because it is virtually impossible to formulate a list of exclusive categories—the sum of the category numbers could be used to identify uniquely the categories selected.

For example, 5 would indicate use of categories 1 and 4, and 15 would indicate use of categories 1, 2, 4, and 8.

3. A "not applicable" category would be provided in each set. There would also be an "other" category with the opportunity for a written insertion. Monitoring of such insertions would reveal when new codes should be added and what they should be. This task would be performed by the Research and Development (R&D) Unit of the SCATT System with the help of National Fellows.

4. The coded preclassification of a document would include (a) its identification by use of an author code and date of entry into the System, (b) specification of the type of document, (c) general characteristics of its subject matter, and (d) specification of the audience to which it is addressed.

If all the documents in the System were not in one language, as we specify as the ideal, then the preclassification would be required to indicate the language of the document.

5. An author's social security number could be used as his identification number. If an author does not have a social security number—for example, in the case of a foreigner—he would have to obtain a number from the National SCATT System, which would maintain a register of such numbers. Similarly, where an organization is the author, it would also obtain an appropriate identification number from the National SCATT System. In cases of multiple authorship only the number of the first author need be used. A digit after the identification number would indicate the number of authors. In this case, however, the document's identification code would also be entered into the register accompanying the social security number of its other authors.

This would enable the identification of the document through any of its authors.

6. A document's date of entry into the System would be indicated by an appropriate code.

7. The categories that would be used to identify the general nature of the document cannot be specified at this time, but they might include the following: book, article, monograph, report, memorandum, letter, pamphlet, raw data, audio or visual tape, patent, and an indication of whether or not the document has been published by a journal or publishing house. The length of the document would also be indicated.

Each of the preceding categories could be further broken down. For example, books could be subclassified as standard, text, collection of articles by the same author, collection of articles by different authors, manual, tables (drawings or maps), and so on.

8. The field(s) of a document would then be indicated by use of codes referring to a classification provided by the National Academy of Science, the National Academy of Engineering, and the National Research and Engineering Councils.

9. This would be followed by a code that indicated the nature of the document. If the document was primarily about one, two, or three other specific documents, these would be indicated by their identification code.

The categories used for indicating the character of the document might include: biographical, bibliographical, theoretical, experimental, conceptual, methodological, how-to-do-it, practice, application, and so on.

10. Finally, a code would be used to indicate the nature of the audience(s) for which the document is intended.

Such a code could include the following categories: specialists in the same field, specialists in other but related fields, specialists in unrelated but scientific or technological fields, and general readers.

11. If a document had identifiable and separable parts (e.g., volumes or chapters) the producer could enter its parts separately as well as the whole. Each entry would require its own abstract, coded classification, index, table of contents, bibliography, and notes. Separate entries should

be made only for parts that can be read independently of the rest of the document.

12. Once the producer of a document has produced a draft of the coded preclassification of his document he would submit it to a Local SCATT Center directly or through a library. The center would then provide him with a list of the ten (or some more appropriate number established experimentally) most recent entries into the System with the same classification. He could then modify his draft classification if he saw fit. If he modified the classification he would repeat the draft submission.

13. A month after a document had been entered into the System five categories relating to its usage would be added to the preclassification scheme. These would be: the number of times the document's secondary description was retrieved over the last month; the number of times the document itself was retrieved over the last month; the three statistics gathered from user feedback (Chapter 2, ¶ 48 and 49). A comparison of the second and first category would indicate the proportion of users who, knowing about the document, wanted it. After a year, five additional categories covering the same statistics over the twelve month period would be used.

Keywords

14. Each document would require selection of ten keywords (or word combinations), or some other number established by research. Each word or expression would have a weight assigned to it (¶ 16). These weights would be expressed to two decimal places, and their sum would equal 1.00.

15. Manual and automated procedures for selection of keywords should be developed and evaluated. Until some one procedure is developed that is demonstrably better than any other, the document producer should have the option of selecting the one he wants to use. However, he would be required to inform the SCATT System of the choice he had made so that its R&D unit could evaluate alternative preclassification and keyword selection procedures by analyzing relationships between the ratio of the number of times a document was found to be relevant to the number of times secondary information about it was retrieved.

This convention could be modified in the light of experience.

16. Indexing services, manual and automated, would be provided at Local SCATT Centers. They would also be provided by private companies.

These companies could use feedback from the System to evaluate their services.

An example of a procedure for selecting keywords follows. Those ten words that appear in the index of the document and that are referred to *most frequently in paragraphs would be selected. (Recall that all documents including articles require numbered paragraphs and indexes.) The weights would be selected as follows: the number (N) of paragraphs cited in the index in connection with the word or expression (i) would be counted ($n_i; i = 1, 2, \ldots, 10$). These would then be summed $n_1 + n_2 + \ldots + n_{10} = N$). The weights ($w_i$) would then be $w_i = n_i/N$.*

17. The producer would be allowed to add keywords to the number specified if he so desired. These additional words would not be used for classification, but they would be recorded and used by the R&D unit when considering the number of keywords that should be required.

18. Once the list of weighted keywords had been prepared it would be submitted (along with the coded preclassification) to a Local SCATT Center or a system-affiliated library. The center or library would provide a list of the ten (or some more appropriate number established experimentally) most recent entries into the system that most closely matched the list of weighted keywords. (The matching process is described below.) The producer could then modify his list and resubmit or submit the original list with his document.

This procedure would serve the purposes of detecting errors in coding and of performing the redundancy check (Chapter 2, ¶ 3).

19. When preparing the set of keywords the producer would have access to a list of all keywords then used in the system. Only words appearing on that list would be allowed, but other keywords could be used to prepare an alternative description, one that is completely unconstrained. Such optional lists would be used by the R&D Unit of the National SCATT Center to study additions to and deletions from the master list. The supplementary description might be used later if either the master list or the permissible length of descriptions was increased.

The imposition of a controlled vocabulary on the coders is necessary in order to limit the complexity of the System and to ensure minimal standardization of document descriptions. Since the terms most frequently cited in a particular index might not belong to the list of authorized keywords, some adjustment of the steps in the coding procedures described in paragraphs 18 and 19 might be required. Thesaurus services could be used to facilitate this process. Computerized, or partly computerized, thesaurus services would be provided by Local SCATT Centers, libraries, and private companies.

20. Lists of documents provided by the SCATT System in response to requests for information would include each document's keywords and their weights as well as the usual identifying information.

21. The R&D Unit of the National SCATT Center would prepare and maintain a manual of instructions on preparation of keyword descriptions.
22. The list of keywords would contain numbered entries for words with multiple meanings.

For example, the word "group," if used in sociology, would have an associated number 1, if used in mathematics, 2, and so on. The user would have to indicate which of these meanings he intended.

RETRIEVING INFORMATION FROM THE SYSTEM

This idealized SCATT System is designed to serve a variety of individuals and organizations with a variety of informational needs and desires. Furthermore, users differ with respect to their personal, organizational, and professional styles. Therefore, an ideal system must provide a variety of ways to retrieve information. The System should be capable of adapting to the ways users usually retrieve information and it should facilitate their learning how to use it effectively.

In the paragraphs that follow we first describe some of the ways that users could devise to obtain information from the System, then we describe several "standard" procedures that would be provided by the System for those users who would not want to design their own searches.

Individually Designed Search Procedures

23. The R&D Unit of the National SCATT Center would periodically publish a description of how its data were organized and how access to them could be obtained. Users would be able to use any of the current programming languages (many of which would be specialized for information retrieval) in designing their own search procedures. They could also use their natural language for such purposes. In the latter case, their algorithms would have to be translated into a machine language either by a SCATT Center or an authorized library, or, if technologically feasible, by automated compilers.

Emphasis would be placed on providing users with easy-to-learn languages for preparation of their own procedures without having to work through a human intermediary—for example, by providing a retrieval language similar to the natural language combined with interactive aids from the computer to resolve ambiguities, diagnose errors, and assist the users in more precisely formulating their requests.

24. A library of standard programs would be made available by Local SCATT Centers. These programs would be modular so they could be easily combined into a larger program. The functions of these programs, their input and output, would be described in a nontechnical language. Whenever options were available, a standard option would be designated. The System would use this option unless the user specified to the contrary.

25. Assistance in designing search procedures and programming them would be provided by Local SCATT Centers, as well as by independent public or private organizations. Users could describe their needs in a functional way and leave design of the search procedure and its programming to specialists.

26. The R&D Unit of the National SCATT Center would provide special files limited in size but organized similarly to the original data bases. These files could be used for experimentation and debugging of programs and also for estimating the cost of running any specified program on the total collection.

27. Users would submit nonstandard programs to SCATT Centers for inspection before use. These would be examined to assure their meeting programming standards. Special testing routines would be developed by the R&D Unit of the SCATT System. Once a program is approved, it would be registered so that it could be used at another time without inspection. The R&D Unit of the System would then have access to these privately developed programs, but other users would have access to them only with the permission of the developer, who could, if he so desired, charge for their use.

Analyses of these privately developed programs by the National SCATT Center's R&D Unit would make it possible to improve continuously the library of standard programs and program modules.

28. Users would be able to extract from the System those parts of the data bases that do not include classified information.

This would enable them to provide retrieval services outside the system. In addition, any independent public or private organization could use this service to develop specialized services to suit a particular group of users.

29. Ready-to-use (standard) retrieval programs would be designed for both repetitive and ad hoc use for the particular purpose at hand.

30. The user could provide the System with his *profile* for repetitive use. This profile would be analogous to a document's description, and thus be expressed by means of a preclassification code or keywords, or both. The user would indicate retrieval procedures and how often they should be carried out.

For example, the user might ask for a list of all new publications in operations research on a monthly basis.

31. For ad hoc searches the user would specify the type of search to be performed in order to receive a list of documents matching his desires.

In general, he would use the program only once; for example, to obtain a bibliography covering the subject of his current interest. Search specifications formulated for repetitive use would probably be more general and broad than those formulated for ad hoc use. Regular users would generally be interested in keeping up to date in an area or discipline that would be rather broad. Furthermore, regular periodic searches would cover only that portion of the literature that had been added since the last search. An ad hoc search would usually be performed relative to a more narrowly defined subject but a longer period of time.

We expect the regular user to employ the preclassification system, but not exclusively, and the ad hoc user to make more extensive use of the opportunities provided by the keyword description of documents. Nevertheless, the modes of retrieving information described below are equally applicable to either type of use, and the user would select the procedure.

32. There would be three basic ways of obtaining access to information in the SCATT System: by use of (a) preclassification, (b) keyword descriptions, and (c) postclassification. These could be combined in many different ways.

Search Procedures Based on Preclassification

33. The user would be able to use any combination of a priori categories to specify the type of documents he desired. He would simply have to fill out a form similar to the one used for categorizing a document. In addition, he could specify the categories that he wanted to *exclude*; for example, he could exclude particular authors, documents that appeared prior to a specified date, or articles that appear in a specified journal. To do so, it would be sufficient to use the sign "—" beside a category. If options to exclude presented to the user were not relevant to him, he would check the option "all."

34. Before carrying out a request (or before doing so for the first time in the case of regular use), its coding would be checked by a Local SCATT Center. An estimate of the cost of its use would be supplied to the user on request. (Such estimates would be provided for all types of search procedures.)

35. Search procedures based on preclassification could fruitfully be used in combination with any other search procedure.

A priori categories would provide a powerful and useful filter; they could exclude documents of a type or types that the user did not want. For example, whatever the particular search strategy chosen, the user might want to obtain only documents that are not older than five years and are designed for specialists in the field.

Searches based on preclassification could also be used to construct a complete bibliography of a given author, or a given institution, and so on.

Search Procedures Based on Keyword Description

36. Users would be free to employ any combination of approved *keywords* to specify characteristics of documents or information to be retrieved.

The user could simply ask for all the documents whose description contain one or more specified keywords. A request for all documents whose keyword descriptions contain "cybernetics" and "control" might be typical.

37. He could also use either the *rank* or *weights* of *keywords* in the description to make his search more precise.

Examples might be: "the documents containing 'cybernetics' in the first rank, and 'control' in the second," "all the documents containing 'cybernetics' and/or 'control' in either of the first two places," "all the documents containing 'cybernetics' and/or 'control' with their total weight exceeding 0.5," and so on.

38. Furthermore, the user could use *decision trees*. That is, he could specify a sequential search procedure each step of which could be based on use of one or more keywords. Which next step would be taken would depend on the results of the preceding step.

In constructing decision trees for search purposes the user could employ several decision criteria; for example, the number of documents in the collection, the proportion of recent (to be specified) documents, their total length, their homogeneity (which could be measured by a similarity index discussed below*), and so on.*

The user would be able to obtain assistance from a Local SCATT Center in constructing his decision tree. For example, it could provide him with a list of keywords closely related to the words selected by him. This list could be obtained by any of the methods for computing the proximity *of keywords discussed below.*

Search Procedures Based on the Postclassification

An unrestricted search based solely on a keyword specification would involve comparing the descriptions of every *document in the System with the specification. There is a need, therefore, to restrict the size of the set of documents to be searched. This need can be filled by use of a statistical filter based on a measure of relevance. Such a filter can be used to create a postclassification procedure. The description of this procedure is provided in a later section. Here we give only a functional description from the user's viewpoint of how this procedure would operate.*

39. The user would formulate his desires by use of weighted keywords much as a document's content is described. Documents within a specified degree of relevance to the user's interest would then be retrieved. (The measure of the degree of relevance, the Similarity Index, is discussed in a later section.)

The user could be assisted in producing a keyword description of his interests in various ways. He could identify several documents that were representative of his interests and have a SCATT computer produce the first keyword description for him. It would do so by determining the dispersion of the specified documents. The measure of dispersion obtained would indicate whether the selected sample of documents adequately represents one or more well-defined concepts. Then the user would have the opportunity to change the sample (by reducing it, for example). The next step would consist of calculating a "mean" keyword description of the sample by use of an algorithm discussed below. This description could serve as the user's tentative profile.

40. After preparing his tentative profile the user would receive a sample of documents corresponding to it for his evaluation. He would also receive an estimate of the average number of documents corresponding to his profile that are entered into the system each month. He could then modify or accept the tentative profile.
41. The request would then be executed. A list of documents corresponding to the user's profile subject to a specified degree of relevance could then be extracted from the System. Whenever he wanted to, the user could modify his profile.

Additional Selection Procedures

The collection of documents obtained from the System after applying any type of search procedure might be too large to suit a user's desires. There would be several ways of limiting the size of the retrieved set of documents, ways that would be independent of the search procedure employed; for example, a random selection of documents. The probability function used to make such a selection could comply with any criterion provided by the user. He could, for example, indicate that he wanted the prob-

ability of choice of a document to be a function of its date of publication. The frequency of use of documents, which would be recorded in the SCATT System, could also be used as a selection criterion or to modify the probability of selecting a document from a list of relevant documents. The user could also impose more complex criteria; for example, not allowing the same author to appear twice on the final list. As with previously described processes, the user could also choose from several preprogrammed procedures.

MEASURING THE PROXIMITY OF TWO KEYWORDS

The Proximity Index *of two keywords would measure the extent to which the two keywords belong to the same universe of discourse. For example, it may be expected that the words "cybernetics" and "control" are very close to each other and that "introversion" and "temperature" are not.*

42. Determination of the proximity of keywords could be based on either the total collection of documents or on a representative sample of that collection. Given any two *keywords,* their Proximity Index is defined as the proportion of document indexes in which both keywords appear. If the two keywords never appear together, their Proximity Index would be 0. If they always appeared together, their Proximity Index would be 1; therefore, for search purposes they could be considered to be synonymous. Use of synonymous words in a keyword description would be wasteful.

Indexes of documents, rather than their keyword descriptions, would be used for computing the proximity of words because they would not be constrained in length; hence they would be more likely to reflect the range of concepts used in documents. The Proximity Index described here, however, does not make use of the number of references to the words involved in a document's index. It would be possible to construct an index that takes these numbers into account, but it is doubtful that the gain thus obtained would justify the additional cost. This, however, is a choice on which further research is needed.

MEASURING THE SIMILARITY OF TWO DOCUMENTS
OR PROFILES

A measure of the similarity *of two keyword descriptions, whether of documents or users, is required for effective search. It is also necessary for construction of a posteriori classes of entries in the SCATT System. In this section we consider the construction of a* Similarity Index (S) *and in the next, the postclassification process based on its use. Several similarity indexes are now in use but not enough is known about them to provide an effective comparative evaluation. Work now in progress at Cornell University (Salton, 1971) is directed at remedying this situation. Output of this work may well require modification of the Similarity Index proposed here.*

Without more research it is difficult, if not impossible, to design an ideal Similarity Index. What is presented here, therefore, is only intended to be suggestive and to stimulate alternative designs which, together with ours, can eventually be evaluated experimentally.

43. The *Similarity Index* we propose would be determined as follows:

43.1. Formulate a matrix (table) in which the words in one description head the rows, and the words in the other head the columns. The words can be listed in any order but subsequent steps may be simplified by ordering them from the most to the least heavily weighted. The weights should be indicated next to each heading.

43.2. Enter into each cell of the matrix the *Proximity Index* of the two words that form the cell. (If the words are the same, the Proximity Index is 1.00.)

43.3. For each cell take the minimum of the weights of the two words forming the cell.

43.4. Multiply the *Proximity Index* in each cell by the result of step 3. (If the words have the same weights, w, and proximity 1 the result will be w. If the two words have zero proximity or if one of the weights is 0, the result will be 0.) Enter this number (x_{ij}) in each cell.

43.5. Now we want to find an arrangement of rows and colums in the matrix that maximizes the sum of elements on the main diagonal. In other words, we want to find that set of cells, one in each row and one in each column, the sum of whose entries (x_{ij}) is maximum. This can be done by use of any algorithm used to solve the Assignment Problem.

43.6. The sum of the entries of these cells gives a *Similarity Index*. This index will equal 1.00 if the words and weights in the two descriptions are the same. Its lowest possible value is 0.

Some examples, using only two-word descriptions are shown in Figure 3-1.

POSTCLASSIFICATION

Organization of the data-base of the idealized SCATT System so as to facilitate search and retrieval would, of course, require a great deal of attention. It is very likely that the same information would be filed in different ways for different purposes. At least one file would be organized around a priori categories and characteristics—for example, date of entry

Words		A	B
	Weights	0.80	0.20
A	0.80	1.00 0.80	0.50 0.10
B	0.20	0.50 0.10	1.00 0.20

$$S = (0.80 + 0.20) = 1.00$$

Words		C	D
	Weights	0.60	0.40
A	0.90	0.70 0.42	1.00 0.40
B	0.10	0.20 0.02	0.60 0.06

$$S = (0.42 + 0.06) = 0.48$$

$$0.50 \; min(0.80, \; 0.20) = (0.50)(0.20) = 0.01$$

Words		A	B
	Weights	0.70	0.30
A	0.70	1.00 0.70	0.50 0.15
B	0.30	0.50 0.15	1.00 0.30

$$S = (0.70 + 0.30) = 1.00$$

Words		A	B
	Weights	0.70	0.30
A	0.80	1.00 0.70	0.50 0.15
B	0.20	0.50 0.10	1.00 0.20

$$S = (0.70 + 0.20) = 0.90$$

Note: The number in the upper left-
hand corner of each cell is
the Proximity Index.

Figure 3-1.   Examples of Computation of Similarity Index (S).

and field. Another could be organized around the Similarity Index of the keyword descriptions of entries. We call classes of documents that are so organized "a posteriori classes" because they are formed after entries are made in the system and they are fitted to the entries rather than conversely.

Our discussion of postclassification is organized around the following questions:

(1)   How would one organize the classes?
(2)   How would one form the classes?
(3)   How would one classify new entries and perform searches?
(4)   How would the System adapt to changes in the nature of its content?

Our discussion of these questions is necessarily general and schematic. Different approaches can be taken to each question. A great deal more research is required before we select the best among them. Therefore, we are more comfortable with discussion of what the ideal System would do than on how it would do it. However, we suggest how the System could do it.

Organization of the Classes

44. The classification system would be hierarchically organized, into different class levels. Each document could belong to more than one of the lowest-level classes, which therefore are not exclusive. Membership would be based on the similarity of the document's and the class's descriptions (see ¶ 46).
45. Classes at higher levels would be organized in much the same way as lower-level classes. They would contain lower-level classes as members.

Therefore, each lower-level class could belong to several higher-level classes, and their measures of association with each could be determined. The optimal number of levels cannot be determined now. It should be determined with full knowledge of the relevant computer programs, the nature of documents in the system, and the ways in which the system is used.

Formation of the Classes

This too is an issue we are not yet equipped to answer. We can only suggest how it might be done. Each step, as well as the whole concept, is subject to modification with further research.

46. The procedure might involve the following steps:
  46.1. A set of document descriptions, referred to as *poles*, would be selected in such a way that the Similarity Index of any two poles would not be higher than a specified level.
  46.2. The similarity of each document with each pole would be determined. A document would be included in the class defined by a pole if its similarity to it was above a specified level. Therefore, every pole would represent a class, and it would constitute the class's keyword description.

Thus, for any document, membership in a given class would take any value ranging from 0 (nonmembership) to 1 (complete membership) (see ¶ 44). In practice, however, the only documents to be considered for membership in a given class would be those having a Similarity Index with that class's pole above a specified level.

47. The variables in this process—for example, the Similarity Index cutoff level—could be modified until the desired number and distribution of class sizes are obtained.

48. Higher-level classes would be formed from lower-level classes, using a sample of lower-level class descriptions as poles. The Similarity Index of each lower-level class to each higher-level class of which it was a part would also be determined. The Similarity Index of each document relative to each higher-level class could also be determined.

Classification of New Entries and Searching in Response to a Request

49. When a new document enters the System it would first be matched with each highest-level class; that is, its Similarity Index to each such class would be determined and it would be included in those in which its index is above the cut-off point. The same process would then be repeated for each next lower-level class contained in the highest-level classes of which it has become a member. This process would continue until it has been placed in the appropriate lowest-level class(es).

50. A similar procedure would be used in carrying out a search on behalf of a user whose request is presented in the form of a weighted keyword description. The user's profile would first be matched with the representative pole of each highest level class. The classes for which the Similarity Index of the user's profile with the class pole falls above a specified cut-off value would be selected. (The cut-off level used would depend on the degree of relevance specified by the user in his request.) This process would then be repeated for successively lower-level classes, using revised cut-off values depending in each case on the requested relevance and on the Similarity Index of that class's pole to the poles of the higher-level classes. This process would continue until the document descriptions to be subjected to a detailed screening were selected.

Adaptation to Changing Content

51. If a new document failed to fall above the cut-off point in at least one class at any level, it would become a new pole.

Thus, old lowest-level classes would be retained but new lowest-level classes would continually be formed. Once formed, old documents in similar lowest-level classes would be considered for membership in them.

Therefore, a document may join additional classes, but it would never be removed from an old one.

52. If the number of documents in a lowest-level class became too large (size to be specified), then this class could be broken into two or more classes of still lower level.

This means that the number of levels used for the classification would change over time and would play an important role in the system's adaptation to the changing content of the literature.

53. The R&D unit would periodically analyze a sample of users' requests. It would isolate from these samples requests that are representative of large groups of users. If one of these requests does not closely match any of the poles, it would become a focus of a new class, as if it were a pole itself.

This procedure would ensure that the classification scheme would not only reflect the structure of the collection of documents, but would also be adaptive to the nature of requests for information.

UPDATING USERS' PROFILES AND
DOCUMENTS' DESCRIPTIONS

In Chapter 2 (¶ 50) the use of feedback features of the system for updating the regular users' profiles was discussed. Here we describe a way of doing so.

54. The number of documents retrieved whose Similarity Index with the user's profile is higher than 0.90, higher than 0.80, and so on, would be determined. The number of documents in each of these categories that were found relevant by the user would be tabulated. If the numbers of documents found to be relevant were not positively correlated with the Similarity Index, this would indicate a need for modification of the user's profile. Such an analysis could be carried out at regular intervals specified by the user or on request. If the correlation were positive then the information used in determining it could be used to evaluate the cutoff point specified by the user in his searches.

55. For each keyword in a user's profile a "utility number," representing the "usefulness" of that particular keyword in retrieval, would be computed as follows. For each of a sample of documents retrieved and found relevant by a user, the Proximity Index of each of the keywords in its description with each keyword in his profile would be computed. The mean of these values for each word in the user's file would constitute its "utility number."

This number could be compared periodically with the weight of a keyword under consideration in the user's profile and would lead to the utility number's modification where necessary.

56. For each batch of documents retrieved and found relevant a "mean" description would be computed. After ten (or some other appropriate number of) such retrievals, the average of these "means" would be determined.

Comparison of these "mean" descriptions with the user's profile would eventually lead to modification of the latter. The reason for introducing a two-stage computational procedure is to avoid storing all of the information about a user's past retrieval.

57. The same algorithms could be used to determine for each document (a) the number of users who have retrieved that document; (b) the number of users who, having retrieved that document, found it relevant; (c) the "utility number" of each keyword in that document's description; and (d) the "mean" profile of users who have retrieved that document and found it relevant.

This information would be fed back to the author(s) of the document, and could be used for eventual modification of its description.

58. Documents inserted randomly into the user's batch of retrieved documents, and found relevant, would have their Similarity Index with the user's profile determined.

This could suggest changing the cutoff value of the Similarity Index between the documents retrieved and user's profile. It could also suggest formation of a new profile for the user.
  Recall that the user would retain absolute control over modification of his profile. The System would only suggest modifications and would only do so on request. It could, however, make such changes automatically if so requested.

59. If a user requested the Local SCATT Center to do so, it would maintain a special electronic file in which all the documents retrieved for him, together with his feedback on them, would be maintained. More sophisticated analyses of the adequacy of a user's profile could then be performed. Similarly, an author of a document could request that the profiles of the

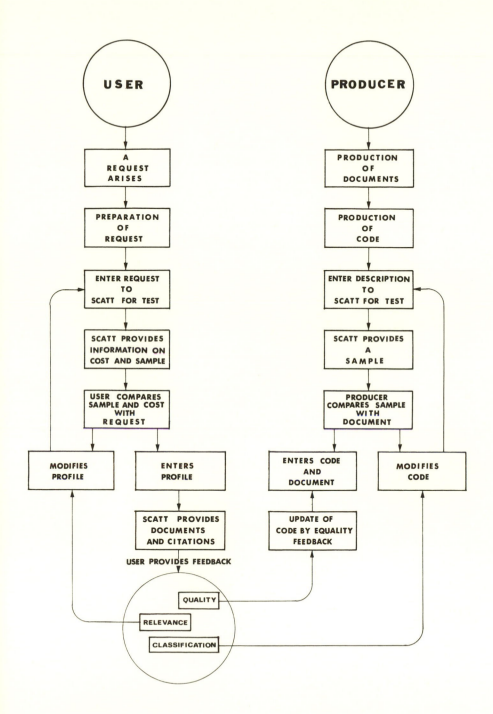

Figure 3-2.   Entry Operations in the SCATT System.

users who have retrieved that document, together with their feedback, be stored in a special electronic file, thereby making it available for sophisticated analyses.

CONCLUSION

The design of the ESR System presented in this chapter was directed at providing the producer with considerable freedom in describing his product, and the user with similar freedom in describing what information he wants and in specifying how the search for it should be conducted. It was also directed at enabling the system, the producers, and the users to learn and improve their interactions with each other. The description of the design is intended to show that these objectives *can* be attained, not how they can *best* be attained. A great deal of research, particularly experimentation, is required before optimal designs can be specified. But we did incorporate into our design properties that would make it possible to conduct such research more effectively than can now be done.

A summary of this chapter may be found in Figure 3-2.

4

TECHNOLOGY
On the Technological Feasibility
of the Idealized Design

INTRODUCTION

Now that the design of the idealized SCATT System and its entry, storage, and retrieval subsystem have been described, we turn to the requirement that this design be technologically feasible. In the following pages we discuss the current technological state and trends relevant to the SCATT System, showing that the design is indeed technologically feasible.

Our discussion is divided into two parts. In the first we consider the functional characteristics of the System, primarily from the user's point of view, with emphasis on man-machine interfaces. In the second part we consider the hardware and software available and imminent that makes it possible to carry out the functions discussed in the first part.

FUNCTIONAL CHARACTERISTICS
OF THE SYSTEM

Use of the System

The principal contact of the user with the SCATT System would take place through terminals that would literally put the System within arm's reach. Keep in mind, however, that human intermediaries would always be available to those who preferred to use them. They would also be within arm's reach, either through the telephone or through an interactive dialogue over a teletype-like terminal.

1. Communications terminals are likely to be readily available to virtually all professional workers and students by the mid-1980s. Most terminals will be used for interactive computing, such as browsing through a bibliographic database. Typically the terminals will have attached keyboards as the primary input device; output will be displayed at keyboard terminals either in hard copy produced by a typewriter-like printer, or in transient form on a cathode ray tube (CRT). If an application requires both simple input (an identification number of a document, say) and simple output (e.g., an abstract of the document), then touch-tone telephones can provide low-cost "universal" access to the System. In such cases, input into the System would be provided through the touch-tone keyboard or (at least in some cases) by spoken word; output would be in the form of an audio-verbal response. Picture telephone terminals may be used to transmit facsimile information and to provide visual information to supplement digital or voice information.

2. The SCATT System would maintain a database containing information about every person and organization authorized to use it. An individual would be described by such information as his name, address, identification number, interest profile, and security codes. Before a person could use the System, authorization data would have to be entered into the database on his behalf. (A large group of users, such as students enrolled in a class, could be given blanket authorization and use a common identification number.)

3. A person wanting to use the SCATT System through a terminal would first have to enter appropriate identifying information. The System would then determine whether that person is an authorized user. If the person's authorization data place any restriction on his access to the System, the System would take appropriate steps to insure that such restrictions are observed.

There is nothing new in providing and limiting remote access to a computer-based system. It is "old hat" in many universities' centralized computing facilities.

4. It is likely that many users would prefer to use the System from their own terminal. If an inquiry was relatively routine, the user could expect to enter his request and obtain a response in a matter of seconds. If the response did not satisfy the person's need, he could then conduct a dialogue with the System. An interactive session of this sort might completely satisfy the user's need for information; if it did, he would sign off and appropriate updating of accounting and traffic records would take place. If the user wanted information that could not be conveniently received on his private terminal (an entire document, for instance), he could initiate the preparation of the output at another site (e.g., a Local

SCATT Center or a university library) for later collection or delivery to him.

5. In addition to private terminals, the SCATT System would include a variety of terminals at SCATT Centers and special user sites such as public and university libraries. These centers would also provide assistance in using the System. They would house specialized terminals that would be uneconomic to maintain privately—for example, high-speed printers, sophisticated graphical display devices, and equipment for generating facsimile microform output. Users could take advantage of these specialized devices either by going to such centers or by initiating output from their private terminals.

We expect that organizations and institutions with large numbers of users of the SCATT System—for example, universities, "think tanks," and corporations with R&D departments—would provide both a number of small dispersed terminals and larger specialized terminals at a central point such as their library or computing center.

6. Although librarians and other information specialists will continue to play an important intermediate role in serving user needs, a strong emphasis would be placed on allowing direct user access to the System. This requires that the design of the SCATT System give considerable attention to human factors. The terminal equipment, languages used for inquiries, display of responses, and response times should be designed to recognize the needs of the user. Many users would be relatively unsophisticated and unfamiliar with the System; the System would provide them with on-line training and user aids. If at any point in the use of the System a person ran into difficulty—for example, not knowing what options were next available to him in formulating a complex inquiry—he would be able to ask the System for detailed guidance (e.g., a list of options, with perhaps an indication of the effectiveness and cost of each). The user experienced with the System could inhibit such detailed instructions, but would always be able to call for them when he encountered an unfamiliar situation. The System would also be able to detect automatically some user problems (e.g., too many erroneous responses) and provide assistance.

We also expect that organizations and institutions containing large numbers of students or practitioners of science and technology would provide instruction on how to use the System. Such instruction would normally be given by someone specially trained for this purpose by the System. Since such instruction would become commonplace in universities, over time knowing how to use the system would become as commonplace as knowing how to use a computer currently is among scientists and technologists.

7. Security of confidential data stored within the SCATT System, and the privacy of users, would be a fundamental design consideration. A user

could obtain complete knowledge of any data retained about him, and have complete control over the outputs he receives. A great deal of attention would be paid to minimizing the possibility that security provisions could be breached accidentally or intentionally. Ironclad safeguards would have to be provided to protect users from possible abuses of their privacy—for example, the disclosure to government agents of a list of readers of "subversive" literature.

Communications

8. Distance between persons and facilities would not impose serious limitations on the SCATT System. All parts of the System would be linked by a communications network, permitting interconnections among persons, databases, and processing equipment. Costs of communication would be largely independent of distance and therefore would not inhibit the exchange of information between distant points.

9. General purpose computer-based networks are likely to become widely available by 1980, linking professional workers in educational and research institutions. A primary motivation for such networks will be the sharing of hardware and (more importantly) programs and databases. The direct exchange of information among users will also be an important objective. The SCATT System would be one of many interconnected networks. The same terminals and communication lines could serve a variety of needs, thus spreading the fixed cost of the terminals and communications network. Central coordinating services—record keeping, billing, specification of network standards, management reporting on the utilization of network resources, dissemination of information about the availability of resources, and the like—would be provided to facilitate sharing of resources across networks. The existence of such services would make it relatively easy to keep track of the use of SCATT documents, to bill for SCATT services, and to make appropriate royalty payments to authors, their institutions, and/or their publishers.

10. The existence of widely used networks would make available a national or international market for a rich variety of SCATT services. Both public and private organizations would thus have a greatly increased incentive to provide services that can be shared from remote locations. This would require added effort on the part of suppliers of services to document and test their services to assure the high quality level needed by remote users. Many services now provided individually by people—such as editing, preparation of art work, and typography—could be provided remotely by specialists serving the national network.

11. A national network would greatly expand the scope of library services. A library would no longer have to depend on its own (or perhaps regional) resources in order to provide material for its users; it could rely on an

automated loan scheme to supply little used or expensive material. This would permit a library to concentrate primarily on relatively inexpensive and popular material, and then depend on external sources for more expensive and less frequently used materials and services. The lending library would recover some of its costs through usage fees. Libraries would thus have a greater incentive to specialize and share their materials.

12. One of the services made available through the network would be an "electronic mailbox." This would permit persons to communicate easily with one another. One person wishing to send a message to another would type an appropriate identifier of the recipient (e.g., name and address or possibly only a unique identifying number). The message would normally be typed immediately at the recipient's terminal. However, if the recipient had informed the System that he did not wish to be interrupted, his messages would be stored temporarily until he asked for his "mail." (The recipient could also permanently bar receipt of specified classes of messages.) The System would permit broadcast distribution of a message to a specified list of persons. It would also allow the maintenance of an interest or expertise profile, so that the sender could transmit a message to an unknown recipient through the designation of the receiver's characteristics. (Again, any person could inhibit the receipt of unwanted messages.) The cost of sending such a message is likely to be considerably less than the cost of a conventional letter; the price paid by the sender, however, would be governed in part by whether or not the message transmitted was subject to the higher rate on unsolicited information.

The "electronic mailbox" could well be to "invisible colleges" what the wheel was to transportation. It would greatly facilitate their formation as well as their operation. These colleges, which we consider to be perhaps the most effective information systems now available, could well move into their "golden age" with the aid of the "electronic mailbox."

13. A system providing electronic mailboxes would greatly facilitate collaborative work among persons who are geographically dispersed. Conferences could be conducted in which each participant received all messages, with automatic identification of the sender (unless anonymity were permitted by participants). All participants could be "on-line" at the same time. Alternatively, it could extend over several days or more, with each participant taking part as his schedule permits. All comments would be recorded and numbered serially, and the System would keep track of messages sent to each participant. A participant could thus receive all past comments since he last participated. He could refer to a message by its serial number if he wished to comment on it. The System would permit retrieval of related comments and those addressed to a specific individual.

14. The electronic mailbox would be ideally suited for communication

between an author and his readers. A reader wishing to comment on a paper could do so merely by appropriately addressing the author (whose address would be printed on the paper as part of its standard documentation). The reader could make his comments anonymously or by name. If an anonymous commentor wished to receive a reply, he could provide a coded designation by which the author could return a reply through the electronic mailbox.

15. An advanced network would permit a wide variety of pricing schemes. Complete information about a service rendered could easily be retained and processed, permitting sophisticated pricing algorithms. For example, prices could be determined on the basis of time of day, priority, long-term versus "spot" rates, or the class of user (e.g., student versus faculty member). Rates could also be based on either the resource inputs used in processing an inquiry (e.g., central processing unit [CPU] time, input/output activity, etc.) or the outputs (e.g., number of citations).

TECHNOLOGICAL BASIS OF SCATT SYSTEM

Electronics

16. The cost of electronic circuits is likely to continue to drop almost to the point of insignificance over the next ten years. According to Withington (1975), by the end of the current decade, a microprocessor capable of providing considerable "intelligence" for a remote terminal (for editing, error detection, etc.) will probably add under $100 to the cost of the terminal (in 1975 dollars). The electronics required to handle more complex functions, such as updating a local database, translating inquiries expressed in a high-level retrieval language, execution of the resulting program, and display of results will probably cost about $1,000. (Other components of a complete system—electro-mechanical devices, software, database creation, and communications—are not likely to experience anything like the rapid technological advances that will be exhibited by microelectronics.)

17. A wide variety of low-cost terminals are likely to be available for use by the SCATT System. The typical terminal will have a keyboard input device attached to either a serial printer (like that of a typewriter) or a CRT display device. The printer will have an output rate of perhaps 30 to 150 characters per second, while the CRT will handle rates of 300 to 1200 characters per second. Compatibility with network standards could be provided by any terminal by means of automatic translation functions built into the communications network. The cost of such terminals will range from a few hundred dollars for the more basic ones to about $2,000 for "intelligent" terminals capable of performing local editing. Their output rates will be modest enough to limit their use to relatively short responses such as citations or abstracts.

18. Microform output devices and high-speed line printers are likely to be capable of speeds in excess of 25,000 lines per minute; optical scanning equipment (to read printed or even handwritten symbols) are also likely to provide very high performance. These input and output devices will, however, remain relatively expensive and therefore would be available only at SCATT Centers or major concentrations of activity, rather than at each work location.

19. Intelligence built into a remote terminal would provide a number of useful features for the SCATT System. Editing, error checking, standard arithmetic operations, assistance in formulating requests or replies, and handling a small local database (for example, on tape cassettes) would all be possible with even a relatively modest intelligence. Local intelligence of this sort could provide greater flexibility, improved response times, and reduced communication costs (since fewer bits would be transmitted to and from a remote computer).

20. Voice communication with a computer is likely to be useful for applications in which input and output are limited in volume and complexity, if for no other reason than that the telephone is so universally available. Computers already can generate output in voice from using a recorded vocabulary or by speech synthesis. On the input side, considerable progress is currently being made in the computer's ability to recognize a limited vocabulary of spoken words.

Communications

21. Communications technology will improve substantially over the next ten years, but probably not to the same extent as electronic circuitry. (Roberts, 1972, estimates that the cost of communications has been declining by a compounded rate of 11 percent per year, whereas the cost of computation has declined at a 56 percent rate.) With these improvements, communication will become less expensive, more reliable, and more varied.

22. A SCATT System would be likely to rely primarily on "packet-switching" communications technology. Each message in a packet-switching network is broken down into separate "packets" of data (usually consisiting of a maximum of 128 to 256 characters). Packets are then transmitted from the sending to the receiving terminal, generally through intervening network nodes. Each node consists of a computer with sufficient memory to store messages while they are being processed or relayed to the next node. Intervening nodes temporarily store the packets while waiting for an available communication line to the next node along the best path to the final destination. If a given communication link is lost through equipment failure, an alternate route is used to by-pass the defective portion of the network. The fragmented message is ultimately reassembled at the final node located nearest to the recipient (for example, a person at a terminal or a computer). All of this would typically take place in about 0.2 seconds. Because of the inherent efficiency of message switching in using communication lines, as well as the power of a stored-program computer at each node, packet-switching networks are able to offer powerful and flexible communication services at low cost. A great deal of redundancy can be built economically into the system so that its reliability would be at least an order of magnitude better than conventional line switching technology (in which a direct link is established from sender to receiver during the entire time span of a transmission).

23. Packet-switching is now available as a common-carrier service from one or more of the so-called "value added" communication companies. The cost for such a communication service (under an already announced tariff filed by Telenet Communications Corporation) is less than $2.00 per hour for a typical interactive terminal session. An extremely important characteristic of this service is that its price depends only on the volume of data transmitted, not on distance. Technological advances, as well as greater competition, are likely to cause rates to fall by 1980, perhaps by a factor of about three (Roberts, 1972). (The rate of cost reduction with packet-switching falls between the rates for computing and communications, which is not too surprising since it combines both technologies.)

24. If communication links in the SCATT network are expected to carry a very high volume of traffic, dedicated leased lines would prove more economical than use of a common-carrier packet-switching network. This would be especially true in the case of high-volume, short-distance links (since the cost of a dedicated line is a function of distance, while packet-switching is not). If traffic warranted it, a private packet-switching network could be implemented and dedicated solely to SCATT traffic (although this is unlikely, given the substantial economies of scale exhibited by communications technology that encourage shared use of common-carrier facilities).

25. Satellite communications technology would have relatively little effect on most of the functions of domestic SCATT System. The cost of long-haul data transmission is typically a relatively small part of the total cost of operating a packet-switching network (less than ten percent, the rest of the cost being switching computers, software development, and so on), and communications costs in general would not be a major component in the operation of the SCATT System. However, when transmitting data over long distances (coast-to-coast or between continents, say) and in very large quantities (such as a large data file, facsimiles, or a book), the very low cost of broadband satellite transmission would be exceedingly attractive. Satellites already offer substantial economies compared to land lines in such applications, and their use is likely to become relatively more attractive as satellite technology matures.

26. Person-to-person voice communication using the conventional tele-

phone network is unlikely to improve dramatically over the next ten years. The widespread use of electronic switching will, however, improve reliability and switching time and increase the variety of services rendered. A much more significant improvement is likely to come from the packet-switching network. Voice messages can be digitalized and compressed for transmission in packets. This may offer a substantial cost reduction in those cases in which the slight delay of packet-switching will be acceptable.

Data Storage and Data Base Organization

27. Data storage technology will continue to make steady progress. A high-performance magnetic-disk device can currently store about a billion characters at a cost per month of less than $10 per million characters. Costs are likely to be reduced by a factor of about five over the next ten years. Very large on-line archival storage devices can currently store about 500 billion characters at a cost of about $0.25 per million characters per month; this is likely to improve in volume and cost by a factor of about fifty by 1985; (access time and transfer rates are likely to improve by a factor of ten over this period—Withington, 1975).

28. SCATT databases would be organized heirarchically. Data frequently called for, needed rapidly, and of relatively low volume would be stored at the top of the hierarchy, in high performance and relatively high-cost devices. Lower levels in the storage hierarchy would consist of lower-performance and lower-cost storage devices. The lower the frequency of access or urgency, or the larger the volume, the lower a given item of data would be stored in the hierarchy. For example, archival data that are infrequently retrieved might be stored on off-line magnetic tape or microfilm. Assignment to a level in the hierarchy would be dynamic. Thus, as data became older and were called for less frequently, they would normally be "bumped" automatically to lower levels. Data could be bumped completely out of the system, but the cost at the lowest storage level is likely to be so low that there would be little advantage to complete "forgetting." An index would be maintained automatically by the System to keep track of the location of each data element.

29. The SCATT database would be geographically dispersed as well as hierarchically organized. Data called for most frequently at a single location might be stored at that site; a user at another location would then have to gain access to the data by means of the communication system using a centralized directory that identifies the storage location of a given type of data. Data needed fairly uniformly at a number of sites would normally be stored at a central point, but if the volume of data were large enough and if the database were relatively static (avoiding frequent updates), it might pay to maintain the database at each site that required frequent access to it. Considerable progress is currently being made in the management of a "distributed" (i.e., dispersed) database, and there appears to be no obstacle to implementation of such a system (see Levin and Morgan, 1975).

Word Processing

30. Word-processing technology will be widely applied in offices and in the publishing industry. Low-cost processors will be available to virtually all professionals for preparation of manuscripts. These processors will provide powerful editing capabilities for adding or deleting words or lines, substituting one word for another, rearranging material, and searching for a specified string of characters. As a byproduct of such word processing, most manuscripts and published documents will be available in machine-readable form.

Word processing technology should greatly reduce the time to publication of accepted manuscripts. It will not have much affect, however, on the time required for refereeing. Recall, however, that where elimination of this delay is critical to an author, he would be able either to enter his product into the system without publication or, in the case of articles, submit it for publication without refereeing.

SCATT Software

31. A SCATT System of the sort envisaged would require extremely sophisticated software as well as powerful and low-cost hardware. Experience with large-scale software development projects has not been very good: they usually take much longer than expected, cost more to implement, and create exceedingly difficult transitional problems. It is important for SCATT to avoid these problems by the use of effective project management techniques. Most successful software projects of any complexity have been implemented with a modular structure and through an evolutionary development. The software of the SCATT System would be developed in this way. We would expect a fairly long development period before the system reached maturity. Emphasis during the early stages would be placed on providing useful interim capabilities that did not preclude future developments. Heavy use should be made of "structured programming" techniques (i.e., modularity, "top-down" development, explicit project management techniques, extensive programmer-support services, etc.). The extensive feedback designed into the SCATT System would provide inputs to the designers, which they could use in modifying the System to make it more efficient and effective.

32. Automatic indexing and abstracting will become increasingly practical as published material becomes available in machine-readable form. Up to now the high cost of converting of documents into machine-readable form has inhibited the automatic generation of indexes, but this is very likely to change by 1980. Sufficient progress has been made in the substantive problem to justify the expectation that automatic indexing will be in widespread use by 1980 (although automatic methods are not likely to produce indexes of the same quality as a competent and experienced person). Automatic abstracting, on the other hand, presents a considerably more complex task. Even here, though, it is likely that the computer can play a useful role (such as identifying sentences or paragraphs with a high frequency of keywords). Such means could be employed by publishers to supplement author-prepared abstracts (See Chapter 2, ¶ 9). Progress has also been made (particularly in the legal field) with systems for retrieval of passages from unindexed material, but such processes would not be common in the SCATT System.

33. Similar advances will take place in information retrieval systems. These will rely on sophisticated methods of data structuring (e.g., permitting links among related documents and index terms). Retrieval languages will be available that provide powerful capabilities for the advanced user (e.g., complex logical search criteria) and extensive aids for the less skilled user. Interactive capabilities will allow the user or information specialist to play an active role in a search process. The system described in Chapter 3 illustrates what will be possible.

34. A number of techniques would be employed to make the SCATT System much more adaptive than most current systems. Some automatic adaptation would be built in, such as dynamic allocation of the storage hierarchy depending on volume of use, changes in priority, and so on. Adaptation through human intervention will be greatly facilitated by modular design of the system, a high degree of parameterization (permitting changes by merely modifying parameters), and very detailed monitoring and (selective) reporting of the System's operation. The design of the SCATT System would be kept largely independent of current hardware, thus permitting the adoption of new hardware without serious impact on the functional characteristics of the System.

35. Although complete security within the SCATT System would be impossible, advances in security techniques should be adequate to ensure a high level of protection (probably much higher than exists in today's informal systems). The system would deny a user access to its resources (data, programs, or communications lines) unless that user had proper authorization. Voice or fingerprint recognition, or the reading of an individually issued identification card, could be used to identify a person who wanted to enter the System. Distinction would be made by the System between those permitted to read a given record, read a given type of data element, change a record or data element, execute a given proprietary program, or modify any of the System's basic parameters or programs (e.g., those that handle communication switching).

For normal use of the System in which "classified" information is not involved, use of an identification card or number might be sufficient. However, if "classified" information were called for, the user would be asked by the System to provide more "positive" identification—for example, voice or fingerprint.

36. Trends in costs heavily favor the substitution of automatic systems for dull, repetitive, labor-intensive tasks. While the cost of computing, storage, and communications have been falling sharply, labor costs have been rising by six to ten percent per year. Thus an automated application that may not be economically justified now may become more attractive with the passage of time. (It may also become attractive because the application is incorporated into a larger system.)

Some argue that the social costs of automation exceed the economic benefits to be obtained from it. This may not be true where there is a surplus of capital and a shortage of labor, but it may be true where the surplus and shortage are reversed. The social costs of automation depend on what it frees people to do (as well as on what it frees them from doing), and on what resources they have to do it with. Therefore, the extent to which the SCATT System ought to be automated depends largely on the social consequences of doing so, as well as on the amount by which it can reduce economic costs. We do not advocate a rush into automation in a less developed country, particularly one whose wealth is inequitably distributed. Even in a well developed "free" society in which there is an abundance of capital but a shortage of jobs, one can legitimately question the desirability of automation where it is not the only way of getting a needed outcome. It should be noted, however, that many of the more powerful features of the SCATT System—"electronic mail," for example— depend utterly on sophisticated technology.

37. One of the most difficult technical problems facing implementors of the SCATT System is dealing with a variety of related networks that serve the user. Proprietary, nonprofit, governmental, and private networks will continue to exist. It is highly desirable to permit these networks to communicate with one another, preferably in a way that does not require the user to learn procedures on a variety of systems.

Standardization of languages and procedures can substantially reduce the interconnection problem; in some cases, however, automatic translation between networks will be a more attractive approach because it will avoid problems of conversion to a single standard.

CONCLUSION

There is nothing in the specifications of the SCATT System that is beyond the technological state of the art. To be sure, the development of the System would call for a high level of technical and managerial skill, and any plan for implementation would have to recognize this. With a suitably cautious but deliberate plan, however, there is every reason to suppose that the System could be implemented from the technological standpoint.

Obstructions to implementation are more likely to be "human" than technological. These are most likely to come from service organizations that have a vested interest in the current system, particularly those that benefit from it. Therefore, we are anxious to involve such organizations in the continuing idealized design process so that a design can be developed that minimizes their opposition and, hopefully, attracts their enthusiastic support. We believe a design is possible in which their opportunities are expanded. We have tried to make this such a design but only through their participation can we determine if and when we have succeeded.

5

ECONOMICS
On the Economics of the Idealized Design

INTRODUCTION

An idealized system ought to be economically viable; that is, capable of sustaining and improving itself. The economics of the system should enable and motivate it to operate efficiently, effectively, and adaptively. The SCATT System can only be self-supporting if it charges for its services. Such charges provide its users with a powerful way of expressing their disapproval of its services: not purchasing them. This, in turn, assures the responsiveness of the System to what the users want.

An additional way of assuring the responsiveness of the System to its users is to require that it compete against other public and private information services which, in turn, have access to all of SCATT's services including its principal data base, the National Document Register. With this, most of the services we have designed into the SCATT System could be provided by others. They could, of course, also provide services that are not provided by the SCATT System.

In the following sections we approach the economic design of the SCATT System by first considering how to finance setting up the System and its subsequent operations. Then we take up the finances of the parts of the System, the centers. Next we consider pricing policies and, finally, the use of subsidies.

SETTING UP THE SYSTEM

A fundamental issue to be resolved is how the SCATT System should be paid for. The design presented in Chapter 2 proposed:

(1) meeting the costs of setting up the System from public funds, and,
(2) thereafter, recovering its total operating costs and generating the capital it requires for investment entirely from those to whom it provides services.

"Setting up" includes more than the investment in plant and equipment; it includes getting over start-up problems, debugging the System, and reaching a volume of activity that can sustain it. The start-up period required to do so would probably take at least a few years.

A set up period would also be required to learn how to price SCATT System services properly. Difficult pricing problems would arise in the early life of the System for several reasons (a) there would be no past experience on which to base decisions; (b) there would be relatively few economies of scale; and (c) it would be important to build up demand as rapidly as possible.

The cost of setting up the SCATT System would obviously be very high. We can see no possible source of such a large investment besides the federal government.

The question remains whether or not the initial investment should subsequently be repaid from operating revenues. We decided against doing so because it would have a major impact on the prices charged for services for a long time. The nature of this impact on charges for services could well be such as to curtail their use and, therefore, to defer significantly the intended social benefits.

We chose the "market mechanism," rather than subsidy, to cover operating costs of the SCATT System for two reasons. First, as noted above, we believe it would make the System more responsive to users' needs and desires. It would provide the user with an instrument (i.e., his payments) with which to draw the System's attention to what he wants. We make it easy for the user *not* to use SCATT by designing it so as to allow, even encourage, competitive systems to operate. Furthermore, the SCATT System would be challenged by competitive information services, and it would similarly challenge them.

Second, and related, is the fact that public subsidy should be based on effective cost-benefit analyses of the System as a whole, or even of most of its parts. Such analyses cannot reasonably be expected because of the variety, dispersion, and subjective and qualitative nature of the benefits to be derived from the System. The market mechanism "decentralizes" cost-benefit analysis and places it in the hands of each user. The overall economic performance of the System, then, would be the result of a large number of such analyses by individual users.

The market mechanism leaves unsolved a very important problem: how should services be provided to those who need and want but cannot afford to pay for them. In an ideal society this might not be a problem, but even an ideal system in a less-than-ideal society must face it. We attempt to do so by encouraging public and private *subsidization of users* of information. It would be possible for public or private sources to provide funds for *use* of the services of either the SCATT System or its competitors. What is precluded in our design, however, is support of the System by other than subsidies to users. This feature preserves the "requirements" that the System become self-supporting and charge for its services and that all those who need them can obtain them.

FINANCIAL STRUCTURE OF CENTERS

1. The National and Regional SCATT Centers would be publicly owned. Local SCATT Centers could be either publicly or privately owned.

Public ownership of the national and regional centers is required to assure a public-service orientation of the System. It will also assure coordination of the local centers and integration of the operations at all levels of the System.
  Private ownership of at least some local centers would be encouraged in order to introduce competition as a stimulus to efficient, effective, and adaptive operations. Privately owned local centers would operate on a "franchise" basis; their franchise could be withdrawn if they fail to perform satisfactorily. On the other hand, publicly owned centers that did not perform satisfactorily could be "sold" to private organizations for operation within the System.

2. There would be a uniform profit-and-loss accounting system in each SCATT Center, whether national, regional, or local; and, if local, whether publicly or privately owned.

Information provided by this system would be used by the National SCATT Centers to set uniform prices for services in all regional and local centers in such a way as to achieve a specified system-wide ratio of revenues to operating costs.

3. If justified by social needs, financial support for some Local SCATT Centers would be provided by their regional centers.

As now designed, the SCATT System would be required to have a local center in every metropolitan area with a population over 225,000, and no less than two such centers per state. Some of them might not be economically justified unless their charges for services were higher than those set by the System. It might be necessary, therefore, to support them with additional funding. Such funding could be provided by the SCATT System without violating the constraint on external subsidies. In addition to this, local authorities and organizations would be able to subsidize local users in ways described below.

4. Whether or not a local center is privately owned would not affect the prices at which it could purchase services from the SCATT System. The only financial difference between publicly and privately owned local centers would be that the latter's operating surplus would be at the disposal of their owners and any loss would be their responsibility.

5. A specified percentage of the annual operating surpluses of all centers, other than those privately owned, would be aggregated and accumulated to meet investment needs. If the accumulated surplus for the System as a whole exceeded costs by more than a certain percentage, the excess would be returned to the U.S. Treasury. (This percentage would be periodically reviewed jointly by the SCATT System and a government agency designated by the Congress.) Congress would also impose an upper limit on the total surplus that the System could make per year. This would encourage the System to reduce the prices of its services if its surplus were large.

6. At least 5 percent of the revenues of the System would be set aside for the funding of research and development.

PRICING

In the following subsections we consider first the pricing of services provided by the SCATT System to producers of information. We also consider fees paid on submission of primary messages for publication outside the SCATT System. We then turn to the pricing of services (other than reproduction of copyrighted material) provided to users of information. The next subsection deals with payments for the reproduction of copyrighted material. Finally we consider services that can be considered to be byproducts of the System.

Services to and Payments by Producers

7. Authors or their publishers would be required to pay a fee to a Local SCATT Center for the redundancy checks that are prerequisite to entry of a document into the System. There would be additional fees for copyrighting documents and entering them into the System. These fees are intended to discourage, but not prohibit, unnecessary publication, particularly multiple publication of the same material even if in modified form.

Note that a portion of the fee collected by the SCATT System for reproducing a copyrighted document would be returned to the copyright holder. This return could therefore, more than pay back his cost of enter-

*ing a document into the System; whether it did so would obviously
depend on the amount of the document's use.*

8. Redundancy checks would not be necessary for commentaries on primary messages but entry fees would.

9. Nondocumentary primary data would require certification by an approved learned or professional society before entry into the SCATT System. Payment by the producer to the society would not be regulated by the system. The cost of entry of the data into the System would be borne either by the society or the author or both.

10. There would be a fee associated with submission of an uninvited to-be-refereed paper to either a journal or a conference. In the event of rejection, it would not be refunded. There would also be a fee associated with submission of an uninvited *not*-to-be-refereed paper. This fee would be higher than that for a refereed paper.

These fees are intended to discourage useless or redundant publication.

11. Recordings of oral presentations made at meetings would be abstracted and coded if entered into the SCATT System. Associated costs would be borne by the organizers of the meeting if the presentation had been invited; they would be borne by the speaker or organizer or shared between them in the case of uninvited papers.

12. An author could, at any time, correct primary or secondary messages that are in the SCATT System. Either the author or the publisher would be required to meet the costs of such corrections.

Services to Users

13. Prices would be related to the costs of providing different grades of service or forms of output. Examples are as follows:

13.1. Service during peak hours would cost more than during off-peak periods.

13.2. Priority service would cost more than nonpriority service.

13.3. Costs would vary with the speed with which output was demanded (turnaround time).

14. Prices charged for *profile-based retrieval* would vary with the costs of the search, depending, for example, on what filters were used, and on the size of the segment of the data-base to be searched.

As was described in Chapter 3, search for documents in response to a user's request would involve two basic steps: (a) selection of poles (classes) that would identify the portion of the data-base to be searched in depth; and (b) detailed search around such poles. Note that in repeti-

tions of profile-based retrievals the selection of poles would not have to be repeated. Only new poles would have to be scanned.

15. The cost of the first step would be much less than that of the second, and it would be roughly the same for all requests. The major cost would be incurred in the second step where it would be necessary to consider each document in the selected sets and compute its Similarity Index with the user's request.

This cost might be reduced, however, if the user specified filters eliminating some types of documents—for example, he might exclude all documents published before a specified date.

16. An estimate of the cost of the second step would be based on the output of the first. It would be a function of the number of classes selected for screening, the number of documents in each of these classes, and the number of documents listed on the output. The last would depend, in turn, on the specificity of filters and, in particular, on the Similarity Index cutoff point. How accurately such estimates could and should be made is a matter that can be determined only through experimentation.

17. Either the estimate would be transmitted to the user for his approval, or, if it did not exceed some amount specified by the user beforehand, the second step would follow without interruption.

Figure 5-1 represents this pricing procedure. The algorithm for pricing profile-based retrieval would add very little to the computational complexity of the System. Furthermore, the price actually charged could be the estimated cost. If this were the case, then comparison of the estimates with the real costs would be made periodically for the purposes of correcting the algorithm used for making the estimates.

18. It would be possible for groups of routine users of profile-based services to economize by adopting a common profile. As noted in Chapter 2, the System could identify users with similar profiles and help them, if they so desired, to develop a common profile.

19. In the case of retrieval based on *keywords* (combinatorial) or *pre-classification*, cost would depend primarily on three factors: (1) the size of the database to be screened, (2) the complexity of the filters, and (3) the number of items in the output. Estimates of the effect of the second and third factors would be provided by the SCATT System based on its past experience. Such estimates might not be precise. Therefore, those who wanted a more accurate estimate could obtain it, for a fee, by making a trial run on a sample of items in a relevant data file.

20. In the case of *nonstandard* (e.g., user-designed) procedures, prices would be based on actual, not estimated, costs unless an "official" estimate had been prepared by the System, at a fee. In such cases the charges

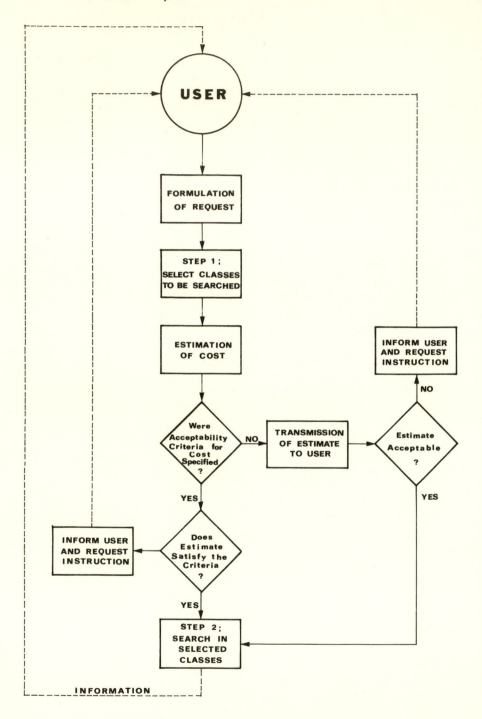

Figure 5-1.   Price Estimation for Profile-Based Retrieval.

could not exceed the estimated cost by more than a specified percentage. Several trial databases would be available to users to enable them to make their own estimates of costs prior to the main run. They would have to pay for doing so.

21. A wide variety of pricing plans would be available. New schemes would continually be tested and introduced so as to suit users' needs and promote better use of the System.

22. Examples of options that would be available to the user are as follows:

22.1. He could obtain services at a lower cost by entering into a long-term contract with the System.

Long-term contracts would reduce the System's risk of underabsorbed fixed capacity costs.

22.2. A collective user could place an upper limit on the costs per unit of time that could be incurred for any or all services, by any or all individual users.

22.3. It would be possible to negotiate special rates for "bulk purchases." This might be done by consortia of users or by organizations with large numbers of individual users.

23. For a fee, any user could have his use of the System monitored and receive advice as to how it might be improved. (This information would be confidential.)

For example, a user who would benefit from entering into a long-term contract because of his heavy use of the System would be notified to this effect. The converse situation would be similarly indicated.

24. As in a normal business, the SCATT System might price new services below their initial cost in order to build up demand for them to a level at which they would become less expensive to provide and at least self-supporting. It could also provide "loss-leaders" or "promotions"; that is, it could offer services that were deliberately underpriced so as to acquaint and thereby attract additional users to these services.

Whenever the System "promoted," however, it would be required to determine whether or not the intended effect occurred. In general, the System would continuously experiment with and monitor the effectiveness of its pricing policies. It would also be subject to a restriction on "predatory pricing" (¶ 29).

Copying Fees

25. The author or publisher would set the royalty that is to be paid either by page or by paragraph whenever a copy is made of a primary message.

It would be desirable to prevent illegal reproduction of copyrighted material. It seems unlikely, however, that purely technological preventive measures will be found since what can be seen by the eye can be "seen" by the camera, if necessary with the use of color filters. Nor does it appear to be sensible to rely upon law alone. Enforcement of a law that prohibits such reproduction appears to be infeasible in a free society.

If the cost of illicitly reproducing a copyrighted document significantly exceeded the cost of purchasing an original, there would not be much of a problem. It may be that this will be achieved through advances in printing processes. It is quite possible, however, that it will not. If this is the case, the following approach seems to us to be one that might work.

26. Photocopying machines would be of two types. One type would be able to produce copies of only single separated sheets (fed through a roller). The other would be able to copy bound documents. A tax would be imposed on each copy made on machines of the second type, except those located in SCATT Centers and authorized libraries affiliated with the SCATT System where copies made of copyrighted documents would require payment of royalties forwarded to the copyright holder. The tax receipts collected from use of other machines of the second type would be paid to the SCATT System and would be used to offset costs of entering copyrighted material into it (thereby benefiting authors and publishers of such material).

27. The tax imposed on bound-volume copying machines would be set high enough to discourage illegal copying of copyrighted documents and low enough to discourage unreasonable prices being charged by publishers of such documents.

28. When a SCATT System or affiliated facility reproduces a copyrighted document it would bind its reproduction in such a way as to make its disassembly difficult and time-consuming, thereby discouraging reproduction of reproductions.

Services to Outside Intermediaries

The SCATT System would provide services to types of customers other than producers and users of scientific and technical information. These would include those who provided information services based on reprocessing information obtained from the SCATT System, those who wanted to make use of the SCATT distribution system to reach particular types of users, and those who wanted to buy information about its users.

29. Information reprocessors would be treated in the same way as other users. The System would not be permitted to discriminate against them in its pricing. Nor would it be allowed to engage in "predatory pricing" in an attempt to take over a market that one of them had developed. The SCATT System would not be permitted to introduce a service already available from another source unless it could profitably do so at a cost to the user that was at least five percent less than that of the service provided by its "competition."

30. Public and private information-service organizations would also be able to market their products through the SCATT System. They would be treated in the same way as the producers of primary information considered above if their product was information that could be stored in the SCATT System. (This would not apply if the product was an interactive service.)

31. Services would be provided at a fee to organizations that wanted to reach users of the SCATT System or to obtain permissible information about them. Such services could not violate the privacy of users nor their requests not to receive unsolicited information through the System. Such services would be priced on a profit-maximizing basis.

32. Miscellaneous services offered by the System—such as the provision of conference rooms, advice on the organization of conferences, and so on— would be priced on the same basis used in pricing information services.

SUBSIDIES

33. Users of the SCATT System, like those of other information services, could be provided with subsidies by various institutions: government agencies, universities, and so on.

Such subsidies would be commonplace where use of scientific and technological information were considered part of the users' jobs or responsibilities, as in the case of students.

34. Examples of the types of subsidy that might be offered are as follows:
    34.1. The simplest form would be a carte blanche. The subsidizer would be billed for the total use of the system by the subsidized user.
    34.2. Another form would consist of providing a fixed sum that the subsidized individual could use in any way he desired.
    34.3. A variation on this would be to meet a given proportion of the costs incurred by the user in question, with or without a limit.

Subsidies could be restricted to specified services—for example, certain users could be subsidized only for noninteractive profile services.

35. Subsidizers could place restrictions on the mode of access to the System—for example, to the priority level for retrieval, the number of uses of the System, the frequency of use, the amount of printed output, the

average total cost per access, and so on. Restrictions could also be imposed on the number of profiles used by the subsidized individuals and on characteristics of these profiles.

The System would facilitate operation of these subsidies by making available to their providers flexible means for control of their use. This assistance, which would serve the System's marketing and promotional objectives, would be provided free of charge.

36. The provider of subsidies could inform the System of the precise nature of a particular subsidy and of the means by which to unambiguously identify those entitled to it. On request, the System would undertake all necessary verification and accounting operations.

37. The System would provide feedback to a provider of subsidies for purposes of control. The nature of such feedback would have to be agreed to in advance by those who are subsidized.

38. At each SCATT Center there would be available services suited to the purposes and needs of actual and potential providers of subsidies. Each center would attempt to improve the range and efficiency of such services so as to minimize the administrative costs incurred by subsidizers.

These services would impose only slight additional costs on the System, since the necessary information processing would be automated and easily integrated into the System's operations.

Subsidies are most likely to be obtained through an employer or an institution of higher learning. This leaves a number of people—such as the retired, those to whom science or technology are hobbies, the self-employed, and precollege students—who might find it difficult to obtain access to the System. For these it would be desirable for the federal, state, and local governments to provide subsidies to be dispensed through public and public-school libraries.

CONCLUSION

The economic design described above indicates how the SCATT System would support itself, how it would price its services, and how users could, if necessary, be subsidized. It would, we believe, motivate the System and its participants to operate efficiently, effectively, and adaptively. The charges that would have to be paid when messages pass from the production to the distribution phase should give authors pause for thought before attempting to rush into print. The profit-and-loss accounting at each center would encourage its responsiveness to local needs, and so on.

The technological design of the System lends itself to such information processing as would be essential for accounting, billing, monitoring, and evaluation of services. Heavy reliance is placed on these last two for, among other things, setting the various thresholds and criteria introduced in this chapter—for example, the level above which income would be returned to the Treasury and the accuracy of the estimates of the costs of associative retrieval.

Before anyone would consider implementing a system of the type presented here, reasonable estimates of the investment required, operating costs, demand, and income of this and alternative designs would have to be prepared. Among the things required before these can be provided are detailed design of the hardware and software of the System and considerable market research. The problems of estimation can probably be reduced by use of a simulation of a mini-SCATT-like System. We plan such a simulation.

Evaluation of the proposed design should not be initiated until it has widespread support from all segments of the community that will be affected by its implementation. Once this is obtained it would be desirable for the evaluation to be carried out by qualified persons who have not been involved in preparing the design.

As we indicated in Chapter 2, estimation of the social benefits of a SCATT System would be very difficult and may be impossible. It is likely, therefore, that the evaluation of the design would be more like that of a proposed free public service. That is, venture analysis may be more appropriate than cost-benefit analysis. It would primarily involve determining whether or not the System would, in fact, be capable of sustaining itself, hence, whether or not its users would sustain it.

6

EXTENSIONS
On the International Connection

INTRODUCTION

Up to this point we have focused on the design of a National SCATT System (NSS) and made only passing references to its interactions with foreign systems. Here we address the subject of international connections: an International SCATT System (ISS). This treatment is the most tentative of any in this report. Its development obviously requires extensive foreign involvement. In the second phase of this project we are engaged in developing such participation.

One objective of an ISS is obviously to facilitate further development of science and technology by enlarging the cast of players as well as by improving communication among those already in the act. But there is also an important related objective: to accelerate the development of the less developed countries. Science, but not technology, is supposed to be a public good, but its distribution among nations is very uneven. The less developed generally cannot afford either to develop their own or to import it, at least not in sufficient quantities to affect their rate of development. Furthermore, although science may be valid transnationally, its applicability is culturally relative. Therefore, it must often be translated before it becomes useful to a less developed nation. The ISS discussed here is intended to improve the distribution of science and to facilitate its translation into culturally relevant forms.

Technology is another matter. Much, if not most, of it is privately owned or held. For that which is not, we seek to do the same through the ISS as we do for science. For that which is privately held, we seek to let the "have nots" know that it exists, what it can and cannot do, and who has it. With such knowledge they can put themselves into a better position to decide whether they need it and, if they need it, whether to try to reinvent or buy it.

Science and technology are increasingly perceived in the Third World as instruments of its exploitation by the First World. An effective ISS could do much to change this perception and the conditions from which it derives.

1. All scientific and technical documents would be produced in, or translated into, one common language.

If and when automatic translation of an acceptable quality becomes feasible, this requirement could be changed. There are, of course, other good reasons not related to SCATT for having a universal common language. Ideally, it would be learned at a very early age when it is easiest to do so. If there were not a language common to all people, then there should be one that would be common to at least all scientists and technologists.

2. Candidacy for a degree in science or technology in any country of the world would require a working knowledge of the common language.
3. Each NSS would be compatible and interconnectable with all others.
4. Every NSS would use the same entry and classification subsystem.

These requirements would, of course, raise fundamental and formidable problems of implementation, but their resolution would be necessary if there is to be an effective ISS. They will be dealt with in a later stage of our research in which we plan to address the question: how could we best move from our current state toward realization of idealized National and International SCATT Systems. One thing is clear: common commitment to an idealized design would facilitate solution of many of the diverse and seemingly intransigent problems of international collaboration.

It is likely that (a) even if a few nations, with more than one native tongue between them, were to initiate an international system, and (b) if this system were to work well, then other nations would join the effort. The task of putting together a small but critical mass may not be very difficult.

The requirements for a common language, system compatibility, and a common entry and classification subsystem are likely to be major issues in any effort toward implementation, but unless these issues are resolved there cannot be an effective ISS. There are, however, other important issues to which the remainder of this chapter is devoted.

PUBLIC WRITTEN COMMUNICATION

This section primarily deals with messages in documentary form together with all associated secondary information. It also deals with the Technology Exhibit and Meeting Information Registers.

Translation and Transliteration

5. Translated documents would include citations of the original versions. There might be delays prior to translation, hence the following requirement:

6. Prior to entry of a document into a SCATT System in other than the common language, it would be abstracted, coded, and given keywords in the common language, and these would be verified before entry.

7. The complete set of characters used in each language would receive a standardized code, establishing a one-to-one correspondence between those characters and one or more symbols in the common language.

Therefore, whenever a proper name, a title, and so on, are transliterated into the common language, their original spellings could be reconstructed precisely.

8. The code of a document would include a designation of the nation in which it had been published or, if unpublished, in which it had been submitted to an NSS.

The two-letter designations currently used on automobile license plates in many countries could, if appropriately extended, serve this purpose.

9. All published documents would initially be entered into the register of the nation in which they were published. All unpublished documents would be entered into the register of the nation in which the author or, in case of multiple authorship, the senior author normally resides. Verification of coding and redundancy checks would be made by the "entering" NSS. However, redundancy checks would cover all documents in the *International Document Register.*

Retrieval from Foreign Systems

10. The ISS would maintain a composite International Document Register that included the content of all National Document Registers. It would also store abstracts of all documents, but it would not store primary messages. These would be available from the National SCATT System in which they were originally entered.

Any nation that so desired could maintain a duplicate of all or part of the International Document Register, or it could use the one in the ISS. A user who wished to interrogate the world's literature could do so by means of this register.

11. A user who wished to interrogate a secondary-information database in a particular foreign country to which his national System does not subscribe could do so, one means being by connection between his NSS and the ISS. Such contact could be in real time. He could also be a registered user of any foreign NSS although this would only be worthwhile in exceptional circumstances.

Although it would be an undesirable obstruction in the way of the free flow of information, it would be possible for governments to impose controls on the purchase of services from foreign SCATT Systems. Such controls might be considered necessary to protect weak balance-of-payment positions.

12. One NSS could register a profile with another and thus receive listings of whatever it considered to be relevant. It could also receive reproducible or nonreproducible copies of documents for appropriate fees.

13. A user who worked abroad could either use the System of the nation he visited or his own NSS with which he could communicate through a Local SCATT Center in the country being visited.

14. All services provided by one NSS to another would be priced in accordance with the principles set forth in Chapter 5. However, the use of one NSS by another could be subsidized by the user's or another government or an international organization (e.g., the United Nations). Subsidization might also be provided for the costs of translation.

Unsolicited Information Distribution Rate

It would be necessary to insure that evasion of the higher unsolicited information distribution (e.g., postal) rate in one country not be possible from abroad, hence the following requirement:

15. By international agreement, the higher distribution rate on unsolicited documents would be imposed by all nations.

Copying

It would also be necessary to insure that piracy of copyrighted documents were not possible and that fees were collected and returned to the author or publisher.

16. Each nation with a SCATT System would use the royalty-tax procedures described in ¶ 25, 26, 27, and 28 of Chapter 5.

Copyright fees could be collected when the copies are made. The fee would be collected by the NSS and transferred to the appropriate country through the ISS. The ISS would serve as a clearing house for all monetary transactions between NSSs.

PUBLIC ORAL COMMUNICATION

17. Taped conference presentations in other than the common language would require for entry in any NSS a written abstract and a coded classification in the common language, which includes a designation of the language used on the tape.

Potential foreign-language users could then indicate their desire for a translation to the author through the international network. This would help authors to determine whether and when to produce such a translation.

18. Experimentation with conferences would include international multilocation events with audiovisual or audiographic hookups.
19. The ISS would maintain in the common language a composite (national and international) Meeting Information Register and Technology Exhibit Register.

This would make them accessible to users in any country. Local and regional meetings would not be so recorded, but access to them could be obtained through the international network.

FELLOWS AND ANNUAL REVIEWS

Some duplication of the fields of responsibility of National Fellows in different countries would be desirable, but too much would be wasteful.

20. Coordination of decisions on fields to be covered by the National Fellows of any country would be a responsibility of the ISS. There would generally be a rotating allocation of fields to countries.

There would be exceptions, of course, where one country "monopolized" a particular area or where others had no competence in it. Hopefully, over time the number of such exceptions would decrease.
    Policy on the design of overlap between countries' responsibilities would be guided by experimentation. Naturally, any country could

appoint a National Fellow for a field that had been assigned to another country. In such cases, several National Fellows working in the same area might well work collaboratively, even in the same location.

ORGANIZATION OF THE ISS

It would be necessary to establish and subject to continuing review those policies and procedures that pertain to the interconnection of NSSs. An idealized design of an organization to perform these and other necessary functions is outlined below. How, if at all, this organization should be related to the United Nations or any other world organization depends on their nature at the relevant time. Since design of such world organizations lies outside the scope of this effort, we have assigned them no role in the ISS other than as a possible source of financial support to less affluent nations. Nevertheless, if the role of any world organization in scientific communication and technology transfer were to change significantly, our design of the organization could easily be adapted to it.

Membership in the ISS

21. Any NSS would be accepted into membership of the ISS provided that it operated in accordance with the criteria established by the ISS. These criteria would incorporate the design features introduced in this chapter or such improvements of them as might be determined subsequently and ratified by its assembly (which is described below).

Responsibilities of the ISS

22. ISS would have the following responsibilities:
   22.1. establishing criteria to which member Systems would have to adhere,
   22.2. providing a forum for discussion of issues and a structure for ratification of decisions concerning the allocation of fields to National Fellows in member nations,
   22.3. undertaking research on international conferences, and
   22.4. interceding with member governments and international agencies on matters that affected any one or combination of its member Systems.

Services Provided by ISS

23. ISS would provide services such as:
   23.1. making available secondary information about and abstracts on world literature,

23.2. running profile-based retrieval similar to that described in Chapter 3,

23.3. organizing or providing facilities for international conferences, exhibits and so on,

23.4. consulting National SCATT Systems on setting up new centers extending the technological base, and so on, and

23.5. administering the financial relationships among National SCATT Systems.

Finance

24. The ISS would be owned by all participating NSSs.

25. Until the ISS became self-supporting, each member system would contribute annually to its financial support an amount proportional to its last gross annual income.

26. The ISS would charge for each of its services and become self-supporting as rapidly as possible.

27. To enforce the payment of levies and bills, the ISS would discontinue service to any member system that had not met its financial obligation within some specified period of time.

ORGANIZATIONAL STRUCTURE

28. NSSs would be grouped into (international) regions.

If all systems were operating in a common language the regions would be organized geographically; if not, region formation would take language as well as geography into account. The organizational relationships of NSSs to (international) regions and the ISS would be similar to those of Local SCATT Centers to (international) regions and the NSS. The differences occur in the governance of the ISS as a whole.

29. The ISS would be governed by an assembly consisting of the directors of each member System. Each would have one vote. Decisions would be made using majority rule. If a group of countries forms one transnational System, the system would have one vote.

The ISS would encourage small nations that could not afford an NSS of their own to form such transnational systems. National centers in such systems would correspond to regional or local centers in an NSS.

30. The assembly would be responsible for all policy decisions, for selection of senior ISS staff members, for establishing membership criteria, for ratifying members, and for expelling those members that did not meet performance standards or financial obligations.

31. The ISS would also have a council consisting of the director of each region (appointed by the Assembly) and its own director general (similarly appointed) who would occupy the council's chair. The council would be advisory to the director general and the assembly.

32. The assembly could delegate some of its powers to committees (for example, for finance or research and development). The council could do the same.

SMALL AND LESS DEVELOPED COUNTRIES

Some nations might be too small or not sufficiently developed to maintain or justify a National SCATT System of their own. As indicated above, they would be encouraged to form transnational systems. The following provisions are intended to supply those who have neither with the knowledge and know-how they require for development.

33. Less-developed countries would be provided with a national terminal connected to the ISS communications network. This would be done with financial assistance from, for example, the United Nations or other sources of foreign aid. The staff required to operate the terminal would be trained by the ISS to provide services within the country. This initial organization would be designed so as to facilitate development into a fully fledged NSS when justified.

34. In addition, each university or approved research institute in these countries would be provided with sufficient assistance to enable it to install a terminal linking it through the national terminal to the ISS.

The communications networks and other services of the ISS and its members provide other possibilities for furthering the scientific and technological development of less-developed countries. Advantage would be taken of these opportunities in the following ways:

35. Either directly through a national SCATT System, or through the ISS, workers in less-developed countries would be able to identify relevant experts who might assist them. It would be possible to contact such experts via their national System by letter, by "electronic mail" or, in real-time, by teletype. Such indirect communication would be necessary whenever the identity of the individual had been obtained from confidential information—for example, his profile. On the other hand, the expert might be a National or Society Fellow whose identity had been obtained from a publicly available register, in which case direct contact would also be possible.

36. It would be possible for organizations and individuals in less-developed countries to obtain tapes of oral presentations made at foreign meetings from National SCATT Systems on payment of a fee. Such fees might be subsidized, and authors (or their assignees) might be prepared to waive their royalties in such cases.

37. It would be possible to organize events at which the author would afterward engage in discussion with the remote audience via audio, audiographic, or audiovisual telecommunications.

38. Another service that would be provided to less-developed countries would be the arrangement of itineraries for their scientists and technologists when on foreign visits. The national System in the country to be visited would be able to design itineraries using information comprised by its Technology Register and its subscribers' profiles. It would be able to check with relevant individuals whether and when they would be prepared to receive a visit from the person or group.

CONCLUSION

Scientists and technologists have, perhaps, come closer to creating truly inter- and trans-national service organizations than have politicians, statesmen, or any other segment of the world's population. Many branches of science and technology have international societies or federations of national societies that serve their members around the world. These societies and federations could play a major role in designing, planning for, and implementing an ISS. By so doing, science and technology may well create a new paradigm for international cooperation and redistribution of wealth, in this case, wealth in the form of information and knowledge. An ISS might well lead to better understanding among nations and inspire extension of their cooperation to other spheres.

7

DESCRIPTION
The Current SCATT "System"

INTRODUCTION

The description of the current SCATT System that follows is not so much a product of our research as a byproduct. It does not provide an exhaustive survey of the literature that deals with the nature of this system. Rather, it attempts to cover just enough of the relevant literature to provide those unfamiliar with the system with a useful overview of it. It is also intended to serve as an introduction and a guide to that literature.

Those who are specialists in the field of scientific and technological information systems will notice that our coverage of the most recent developments and developments still going on is quite incomplete. There are two reasons for this. First, many of these developments are known only to a small group of specialists, often only to those working on the development itself. Second, an exhaustive survey of the current state of the relevant art and science would be possible only if the kind of SCATT System we have presented here were available. Currently, it is extremely difficult, if not impossible, to identify, let alone acquire, all the relevant literature.

Before proceeding with the description it may be helpful to recall a few concepts and distinctions that were used in Chapter 2 and which are called back into use here. We divide communications into formal and informal, the formal tending to be asymmetrical (one-way) and prepared and the informal, symmetrical (two-way) and spontaneous. Since messages can be either auditory or visual, we identified four types of communication: formal auditory, informal auditory, formal visual, and informal visual. We focus on formal communication rather than informal because the system does, but we recognize the extremely important role of informal communication.

Communication that involves recorded information can be disassembled into four functionally defined stages: production, dissemination, acquisition, and use. These functions are interdependent; an individual or organization may be involved in several or all of them at the same time. Nevertheless, most of the subsystems of the current scientific communication and technology transfer system are primarily associated with one of these functions. Therefore, we consider these subsystems in connection with their principal function but try not to overlook their involvement in the others.

Our treatment of the current system is not purely descriptive. It is interspersed with evaluative comments that we have set off so they can be identified easily. We call attention to what appear to us to be deficiencies as they arise, and we bring them together in section summaries. Our "in-process" comments are set off in italicized paragraphs to facilitate their identification.

PRODUCTION

The following steps are often involved in preparing a message for publication:

(1) preparation of an outline or notes,
(2) informal discussion of the ideas involved with colleagues,
(3) preparation and circulation of a draft in-house or outside to friends,
(4) discussion of content at small, informal meetings,
(5) preparation and presentation of a paper at a formal meeting, and
(6) preparation of a final version and submission for publication.

Whatever steps are taken, they usually proceed from less to more formal communication. An author is generally more interested in feedback in the early stages than he is in later ones. The steps are much the same in different fields, but their order and duration and the use of media in them vary.

Estimates of the average time consumed by each step leading to publication were made by The Center for Research in Communication at Johns Hopkins University (Garvey, Lin, and Nelson, 1970). These estimates, shown in Figure 7-1, were obtained from its continuing survey of more than 30,000 participants in the physical, engineering, and social sciences. The center noted that the dissemination of information in the physical sciences starts with presentations to the most specific audiences and ends with the most general. This process takes 80 percent longer in

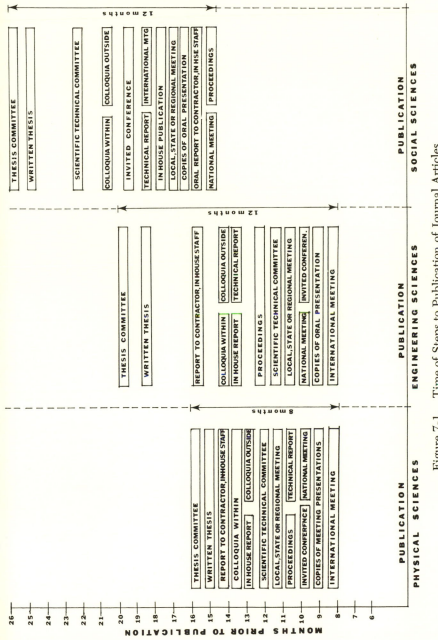

Figure 7-1. Time of Steps to Publication of Journal Articles

Source: Reprinted by permission of the publisher, from *Communication Among Scientists and Engineers*, edited by Carnot E. Nelson and Donald K. Pollock (Lexington, Mass.: D. C. Heath and Company, 1970), p. 73.

the social sciences in which the presentations do not move as rigidly from specific to general audiences. The center suggests that this may be why it takes longer to disseminate information in these sciences.

Garvey and Griffith (1972) flow-charted the document production process of members of the American Psychological Association (Figure 7-2). The time from initiation to publication of work in psychology is not very different from that in the social sciences. The steps are essentially the same but the order differs a bit. Like physical scientists, psychologists first report informally to colleagues in their own institutions and gradually increase the size and formality of their audiences.

This seems to suggest that the order of the steps taken by social scientists, contrary to the suggestion of the Johns Hopkins Center, is not responsible for the longer time from initiation to publication taken by the social sciences as compared with the physical sciences.

It is apparent that the major delay in the production process occurs between submission of an article or book and its publication. This delay is due to the time taken by refereeing, typesetting, proofreading, and printing.

Reductions of lost time from these sources should be a major objective of any system redesign because such delay can be very costly to both science and society. New technological developments (discussed in Chapter 4) are reducing this delay.

The time from submission of a manuscript of an article to a journal to its publication varies considerably within and between fields. In the physical and engineering sciences approximately 8 months are required (Figure 7-1). In the biological and biomedical area the average processing time is 6 to 7 months (Orr and Leeds, 1964). The American Psychological Association (APA) (1963) reports: "The authors of 396 articles in 25 journals publishing articles of importance to the field of psychology were studied . . ." Manuscripts submitted to APA journals take about 15 months to publication. For non-APA journals, however, the average processing time was only slightly more than 6 months (APA 1963:180-181). Publication processing time in the social sciences is also about 15 months. The average delay for *Operations Research* is 15.8 months; for *Management Science Theory*, 21.9 months; and for *Management Science-Application*, 17.1 months (Bach, 1974).

Garvey, Lin, and Tomita (1972) found that between submission and publication authors usually refrain from reporting the work involved but start new work that either stems from or is built upon the work reported in the manuscript. "That authors ceased reporting their previous work about 7 months before its publication (or shortly after submission of their

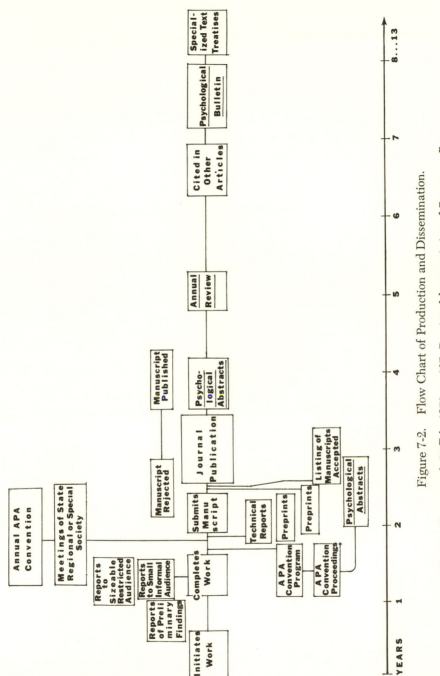

Figure 7-2.   Flow Chart of Production and Dissemination.

Source: Garvey and Griffith, 1972, p. 127. Reprinted by permission of Pergamon Press.

manuscripts) is not surprising, since most of them were well into producing new information which in some respects would render the previous work obsolete" (Garvey, Lin, & Tomita, 1972:219). In order to accelerate dissemination of their work, authors in some fields increasingly resort to formats other than articles—in particular, to technical letters, listings of titles of forthcoming papers, and advanced abstracts. In the *Journal of the American Chemical Society* the proportion of short communications increased from 18 to 46 percent in a nine-year period (Gannett, 1973).

Rejection rates vary among fields. In a study conducted by Garvey, Lin and Tomita (1972), which covered 3,676 articles, it was found that on the average one out of eight articles had been previously submitted to another journal. Focusing on 434 of the articles covered they found that seven out of ten had been rejected essentially without comments, the others with comments. A large proportion of the rejected articles were submitted to another journal, but only half of the authors had made any changes or revisions before resubmission. Most (nine out of ten) were accepted by the second journal. Six manuscripts had been submitted to four or more journals. Each rejection delayed publication from 3 to 3½ months.

Garvey, Lin, Nelson, and Tomita (1970) cite rejection rates of 6 percent for optical scientists and 45 percent for sociologists. In another study covering eighty-three journals, rejection rates varied from approximately 10 to 80 percent. These rates were found to be lowest in the physical, chemical, and biological sciences; higher in statistics, mathematics, and the social and behavioral sciences; and highest in the humanities. According to Kochen and Tagracozzo (1974:204), "Not only is the rate of rejection higher in the 'soft' as opposed to the 'hard' science journals, but also in the interdisciplinary journals as compared to specialized journals."

Preprints and presentations are used widely to obtain feedback before publication, depending partially on the importance of the effort. Garvey, Lin, Nelson, and Tomita (1972:166-167) note:

As the process of scientific communication has grown longer and more complicated in recent years, the national meeting has developed a distinct and increasingly important function in the overall communication process. Nowadays the national meeting is integrated into this total process in such a way as to usually constitute both the first major occasion for early dissemination of scientific work and the last major informal medium before such work becomes temporarily obscured from the public during the relatively long period between first submission of the manuscript and their eventual journal publication.

National meetings provide an author with visibility. Therefore, many younger scientists make presentations to gain entrance into an informal network or "invisible college." Doing so also provides recognition "back home." Younger scientists use these meetings also to acquire information about their fields and those who work in them and to find jobs. Well informed "old timers" use meetings to meet and "catch up" with old and less accessible friends. National meetings aggregate a large proportion of the information produced annually by a discipline, usually considerably more than can be assimilated at the meetings themselves.

Although many young scientists and technologists go to such meetings in the hope of finding others who share their interests, the meetings are conducted in such a way as to make doing so difficult. Furthermore, where such contacts are made, the meetings minimize the possibility of extended informal discussion among those who have found each other. Time and facilities for informal discussion are usually severely limited and opportunities for extended discussion at formal sessions are very rare. Moreover, they usually provide insufficient time for complete presentations; most speakers are restricted to verbal digests.

Although experimentation with the form and content of meetings is continuous, most of them adhere to the conventional format. Here too there is considerable room for improvement.

Garvey and Griffith (1972) note that the preprint is used increasingly as a way of attracting comment before submission for publication as well as for accelerating dissemination after submission. Lin, Garvey, and Nelson (1970:57) found that preprints are usually sent to interested colleagues, others with whom the author would like to communicate, and those who have previously requested reprints.

Producers of scientific and technological messages are, of course, among the heaviest users of messages from others. Hence, *use* is an important part of production. Message producers make considerable use of the informal networks that have come to be known as *invisible* or *hidden* colleges. Although scientists use formal channels of communication more heavily than technologists do, both use informal channels equally. Invisible colleges of scientists tend to be more dispersed than those of technologists for two reasons: (a) scientists working on the same subject tend to be more dispersed and (b) scientific research is less likely to be "classified" whether in the public or private domain. In general, there are fewer barriers to communication among scientists who are in different organizations than among similarly separated technologists.

In technology, message producers depend most on others working in the same organization. Allen (1970:20) found that communication within technologically based organizations has a significant effect on productivity and the quality of the work produced. Therefore, he argues, organizations should be designed to promote such communication. He and Fusfeld (1974) found that the physical environment and its organization can also affect the interaction of researchers. According to Lin and Garvey (1972:15), "In general, communication is found to be more frequent and effective when the physical distance between the participants is small and when the participants consider a distance to be appropriate for the discussion of the specific topic."

This suggests that researchers should have more freedom in locating them-selves physically within the organizations of which they are part and that they should have more opportunity to rearrange themselves as their needs and interests change.

Scientists use their colleagues as a source of information and criticism. Lin, Garvey, and Nelson (1970) found that at some stages of their work feedback from others modifies a scientist's current or planned research, changes its directions or scope, or introduces new findings of others to it.

Informal networks have been receiving increased attention from com-munication researchers. The American Psychological Association (1969: 253) reports:

A high degree of organization is associated with a) a limited number of institu-tions having research facilities . . . , b) a single specialized organization con-taining most researchers in the field, c) many student-teacher relationships . . . , d) long-term commitments to research in the area, e) the area being the prin-cipal research interest of most researchers.

A small highly technical field . . . seems likely to have a high level of informal exchanges but not necessarily a high degree of social organization.

The absence of both communication and organization can probably occur only in a field . . . which most researchers regard as a subsidiary interest and which does not generate its own conceptual framework.

Diana Crane (1972) believes she found a relationship between the development of knowledge and the use of informal networks (see Figure 7-3). She notes that there is little organization of a field until a paradigm appears in it. According to Griffith and Miller (1970) as paradigms develop and while those involved believe their work is new and signifi-cant, they tend to form highly coherent groups. When major solutions are found and anomalies begin to appear, specialization as well as contro-versy increases. This ultimately leads to a decline in the membership of the group (Crane, 1972).

In every highly structured field there are a few key people who serve as "gatekeepers" (Allen, Peipmeir, and Cooney, 1972). Most researchers rely on them for relevant information about what is going on in the "out-side" world. Susan Crawford (1971) found that in a field of 218 sleep researchers the informal network included 165 of them. The key people in the network had more contacts with other scientists, had a higher level of productivity, were cited more frequently, and were read more widely than the others. Ninety-five percent of the members of the network were within two steps from a member of the core and 61 percent were in direct

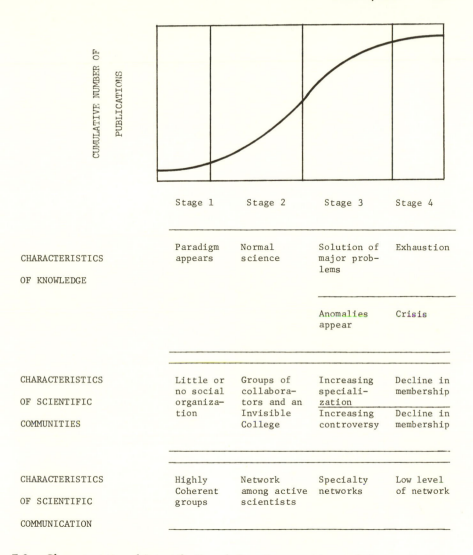

Figure 7-3. Characteristics of Scientific Knowledge, Communities, and Communication.

Source: Adapted from Diana Crane, *Invisible Colleges: Diffusion of Knowledge in Scientific Communities*. Chicago: University of Chicago Press, 1972, p. 172. Copyright © 1972 by the University of Chicago.

contact with one or more of them. Hence, she observed, members of the core are "highly influential in the field" and are "in a position to monitor new information and in likelihood exert strong control over type of research and funding" (Crawford, 1971:309). Ronald Havelock (1971) found similar characteristics among highway safety engineers in urban centers.

Many feel that the "invisible college" is the most effective and satisfying way to obtain the kind of information, constructive criticism, and moral support that most producers need. A member of such a network has relatively complete control over its use and can modify his participation in it when he sees fit. But development of such a network takes time and, once established, tends to add new members slowly. Younger scientists and technologists do not have as easy access to them as do those who are more mature unless they have made a major contribution to their fields. In a sense, invisible colleges are most available to those who need them least. Their memberships seldom bridge generations; they tend to include those who "grew up together." They remain fairly small to facilitate communication among their members.

There is a need for developing ways by which younger scientists and technologists can become part of existing networks or can form new ones more easily and rapidly. Old ones may have to become larger than they currently are, but by use of modern communication technology this can probably be done without overloading the core members. Hidden colleges are seldom spontaneously or deliberately formed. It is worth considering the extent to which their number, if not their effectiveness, might be increased by design and through deliberation.

Only a small percentage of research and development is carried out in an academic environment. Most of it is conducted within industry where much of the information generated is treated as proprietary (see Table 7-1). Some such information is transmitted through informal networks, but such networks are usually least developed and dispersed among researchers working in industrial environments. (Some information, of course, is transmitted by scientists and technologists moving from one

Table 7-1.

Research and Development Expenditures

	Total R & D		Total Research Basic & Applied		Development	
	1971	1972	1971	1972	1971	1972
Total	27,315	29,150	10,167	10,802	17,148	18,348
Federal Government	4,153	4,480	2,012	2,157	2,141	2,323
Industry	18,314	19,540	4,013	4,205	14,301	15,335
Universities & Colleges	3,070	3,280	2,935	3,154	135	126
Universities & Colleges Associated Fed. Funded R & D Centers	716	769	480	515	236	254
Other Non-Profit	1,062	1,081	727	771	335	310

Source: U.S. Bureau of the Census, 1973.

such organization to another.) There is a widely perceived need for disseminating much more of the information and knowledge unnecessarily withheld by public and private organizations.

Up to this point our discussion of the production of messages has focused on preparation of journal articles and papers for presentation at meetings. There are, of course, other types of messages. Some of the more common of these and the media they characteristically use are shown in Table 7-2.

Much of message production, particularly oral, is not for publication and is quite informal. In a study of 1,500 chemists, Halbert and Ackoff (1958) reported that scientists in their sample spent almost 50 percent of their time at work in communication, a little more than three quarters of which was scientific and a little less than one quarter of which was business-related. Tables 7-3, 7-4, and 7-5 show some of their results. Note that chemists spend almost 10 percent of their time "sending" messages either orally or in writing, and almost 20 percent of their time in discussion. These data reveal that only a small portion of their message production is for publication or formal presentation.

The time and effort required to publish are clearly deterrents to publication. Fortunately, they discourage publication of much that we can "do without." But unfortunately they also discourage publication of much that we could "do better with." To be sure, there is considerable need for better filtration of messages heading for publication and greater condensation of what is published, but there is also a great need for accelerating and facilitating the publication of good material. Much more research than has been carried out to date is required before these needs can be effectively satisfied.

Table 7-2.

Classification of Scientific Messages

	Formal	*Informal*
PRIMARY	Journal articles, monographs, dissertations, proceedings of meetings, reports, preprints, patents	Meetings, seminars, letters, conversations
SECONDARY	Bibliographies, listings, indexes, abstract journals, SDI services, dictionaries, catalogs, newsletters	Letters, conversations
TERTIARY	Critical reviews, state-of-the-art surveys, review books, yearbooks, handbooks, textbooks, encyclopedias	Lectures (classes), conferences, symposia, letters, conversations

Table 7-3.

How Chemists Use Their Time at Work

| | | Percent of Time | |
Activity	Minimum	Average	Maximum
Scientific communication	15.7	33.4	61.4
Business communication	0.8	10.4	40.0
Thinking or planning alone	0.0	6.0	25.6
Equipment setup	0.0	6.2	25.9
Equipment use	0.0	23.4	70.1
Data treatment	0.0	6.4	31.6
Personal and social activities	0.0	9.8	33.6
Miscellaneous	0.0	4.4	13.6

Source: Halbert and Ackoff, 1958, p. 96. Reproduced with permission of the National Academy of Sciences.

Table 7-4.

**Percentage of Chemists' Time Spent in Sending
or Receiving Written or Oral Communication**

	Sending	Receiving	Total
Written	5.0	7.2	12.2
Oral	4.5	3.8	8.3
Total	9.5	11.0	20.5

Source: Halbert and Ackoff, 1958, p. 96. Reproduced with permission of the National Academy of Sciences.

The outputs of scientific research—information, knowledge, and under-standing—are normally represented in messages, hence they are usually consummated in documents. The outputs of technological development are normally products or processes that may or may not be represented in documents. Of those documents that are produced by technologists, many are classified or are otherwise withheld from the open literature.

The proliferation of scholarly and scientific journals is currently a source of widespread concern. In an article about a related study to be conducted under the auspices of the American Council of Learned Societies, *The New York Times* (June 29, 1975) mentioned that an estimated 10,000 such journals are now published in this country. E. E. Booker, who is to lead the study, is reported to have said that the "publish or perish" syndrome "drives young scholars to publish trivia." Another point made in the

Table 7-5.

Breakdown of Chemists' Scientific Communication

Activity	Percent of Time		
	Minimum	*Average*	*Maximum*
Total Scientific communication	15.7	33.4	61.4
General discussion	0.0	10.3	35.3
Oral, non-discussion	0.0	9.2	28.0
Total			
Written	3.9	14.3	45.0
Unpublished written	0.0	9.5	40.0
Published written	0.0	4.9	18.4
Sending, oral	0.0	4.5	17.7
Receiving, oral	0.0	3.8	19.4
Sending, written	0.0	5.0	15.0
Receiving, written	0.0	7.2	18.4
Retransmittal	0.0	2.7	20.6
Reading articles	0.0	2.6	13.7
Reading for use	0.0	3.9	14.3
Reading for general information	0.0	3.2	18.4
Communication with other scientists, non-chemists	0.0	2.7	16.3
Communication with other company personnel, secretaries, technicians	0.0	7.1	25.7
Communication with chemists	5.3	21.4	54.5

Source: Halbert and Ackoff, 1958, p. 96. Reproduced with permission of the National Academy of Sciences.

article is that many scholars believe that studies in this area have had little value.

Producers of technology use both the open and the closed literature much as scientists do, but they appear to make less use of the open and more of the closed. As previously noted, technologists tend to depend more on resources and sources within their own organizations than scientists do. Thus, the transfer of technology involves the transfer of people, products, and processes more than scientific communication does. However, it also involves transmission of recorded messages, particularly technical reports and patents, in the open literature. According to the U.S. Bureau of the Census (1973) there were 105,300 patent applications in 1972. The differences between scientific communication and technology transfer are reflected in the fact that scientists in underdeveloped countries have better access to the science than their technologists have to the technology of developed countries.

This state of affairs derives from the fact that more of technology than of science is treated as nonpublic property. The ability of individuals and organizations, public or private, to benefit competitively from ownership of technology has undoubtedly stimulated much of both public and private investment in it, but restrictions on dissemination of its output have equally clearly retarded technological development. We do not know the net effect of these opposed forces nor how to determine it. In the absence of relevant information, opinions on the subject vary widely. Many agree, however, that more information is withheld from the open literature than is necessary. This is not surprising. To those who invest in the development of technology, the release of information that they should have retained is considered a loss to them; the retention of information they should have released is, or is perceived as being, a loss only to others.

The withholding of technology and, to a lesser extent, of science has produced several practices, the morality, if not the legality, of which is being questioned publicly. First, it is common practice for private organizations and national governments to try to lure away technologists employed by competitors so as to get access to their private information. Secondly, espionage is widely used by both public and private organizations.

The conflicts between public and private interests in science and technology, as well as in other areas, do not seem to have a general solution. We seem destined to continue to deal with them on an ad hoc *basis, in the specific situations in which they arise. This, however, does not preclude improvement in how we do so.*

DISSEMINATION

Dissemination includes all processes by which messages or other scientific and technological products (a) are transmitted from the originator (distribution) and (b) are called to the attention of potential users (referral).

In this section we are primarily concerned with secondary and tertiary messages by which attention is called to primary messages. But first we consider the distribution of primary output.

Distribution

Formal or informal messages may be sent directly from a producer to a potential user as in writing a letter and sending a preprint, print, or reprint. They can also be sent by the publisher either by request (e.g., subscription) or without it (unsolicited). They are displayed and distributed by intermediaries such as libraries, bookstores, or friends.

Our studies of a small number of scientists show that a high proportion of the documents and other messages that they receive are unsolicited.

This suggests the need for filtering out irrelevant information. Every well-known scientist we interviewed said that the amount of material received is more than can feasibly be read. This suggests that condensation is also important.

Outputs of science and especially technology not in the form of messages may also be distributed by the producer, "publisher," intermediaries, or other users. Products and processes are frequently displayed, exhibited, or demonstrated at meetings, conventions, and museums, or are observed where they are used.

Referral

In this section we first consider formal referral processes and then various types of services that carry them out.

Referrals are couched in a number of forms—for example, citations, bibliographies, listings, indexes, catalogs, reprocessing (fitted indexes), abstracts, reviews, syntheses, and combinations of these.

Citations are elementary secondary messages. If the primary message referred to is a book, citations usually identify author(s), title, publisher, publisher's location, date of publication, and sometimes number of pages and illustrations and price. In the case of journal articles, the journal title, volume, number, date, and beginning and ending pages of the article are usually provided.

A *bibliography*, according to Thelma Freidas (1973:132), is a "representation of a body of literature in a compressed, abbreviated form" and is "any publication that is intended primarily to point out where to go for information, rather then to convey information directly." Bibliographies may appear either as part of a primary message (article or book) or as a separate article or book. The items cited in a bibliography normally share some contextual property.

A *listing* is a bibliography of items that have some common property other than their content—for example, recent publications or publications of a single author or publisher. Journals often provide listings of the contents of related journals or of books recently received. There are services that announce available translations, such as *The Guide to Scientific and Technical Journals in Translation* and the *Comprehensive Guide to Scientific and Technical Translations*, both published by the Special Libraries Association (SATCOM, 1969). *Current Contents*, published by the Institute for Scientific Information (ISI), consists of reprinted tables of contents. It is issued weekly in six editions:

Agriculture, Biology, and Environmental Sciences
Social and Behavioral Sciences
Clinical Practice

Engineering and Technology
Life Science
Physical and Chemical Sciences

Each edition covers approximately 1,000 journals.

Indexes are bibliographies organized around subject matter and/or authors. "Within the framework of some classification, indexing, or other content-description scheme, they provide bibliographic data, generally on journal articles, as to the authorship, article and journal titles, volume, issue, page location, and year of the recorded texts to which the index makes reference" (Weinstock, 1975:143). According to the Committee of Scientific and Technical Communication (SATCOM) (1969:140) of the National Academy of Sciences–National Academy of Engineers, there are four principal types of index:

(1) "Descriptive cataloguing indexes . . . that use elements of the basic bibliographic record" (e.g., Chemical Abstracts' *Author Index*, BIOSIS's *Author Index*).

(2) "Alphabetical subject indexes in which references are entered under subject/content/terms" (e.g., Chemical Abstracts' *Subject Index, Index Medicus*, and Association of Computing Machinery's *Subject Index*).

(3) "Keyword-in-context (KWIC), keyword-out-of-context (KWOC), and keyword-and-context (KWAC) indexes, in which keywords of a title appear as index terms accompanied by contextual matter and a key to a full citation" (e.g., *Dissertation Abstracts'* Index, Chemical Abstracts' *Keyword-In-Context Issue, Chemical Titles* KWIC Index, ISI's *Permuterm Subject Index*, and BIOSIS's *Biological Abstracts Subjects in Context* (BASIC), *Bibliography of Agriculture*).

(4) Citation indexes, in which citations of publications that have referred to a particular publication are grouped with a citation of that particular publication (e.g., *Social Science Index, Science Citation Index*, and *Shepard's Citations in Legal Literature*).

The *Science Citation Index*, published by the Institute for Scientific Information, covers 2,400 journals. At the end of 1971 its data base contained more than 27 million references to 10 million published items. The "top" 152 journals accounted for more than 50 percent of the journal references, the top 767 for 75 percent, and the top 2000 for 85 percent. (This reflects the uneven use of journals, to which we make reference below.) A paper was cited an average of 1.7 times per year. It was most frequently cited in the two years following its publication.

Wade (1975) gives a short review of the citation index as an indication of scientific activity. It seems natural to assume that articles that are frequently cited are likely to be "better" on the average than those that are cited less frequently. But there are two studies that contradict this assumption. Brodman (1944) obtained a ranking of periodicals in order of value as judged by members of the Department of Physiology, Columbia University, and another ranking in order of number of citations from some of the leading journals in the field. She found such a low correlation that "a grave doubt was thrown on the validity of the . . . basic assumption." Therefore, she concluded that little dependence could be placed on citations as a guide to the value of a periodical. Martin (1963:102) reports on a similar study applied to articles in the field of operations research:

We took the opinion of experts as a measure of value and asked whether high-citation articles tended to be those which experts agreed were the more valuable ones. Our results indicate that experts choose ordinary-citation articles almost as frequently as high-citation articles as being the more valuable.

Martin also asked the experts who had evaluated the sample of articles used in his experiment what criteria they had used. The five most frequently cited criteria were:

(1) the amount of inspiration it gives to others working in the field,
(2) the quality of the research done,
(3) the number of applications that could be made from the results reported,
(4) the amount of synthesis of known facts, provided it is carefully done, and
(5) the effect of the article on general principles of the discipline.

Co-citation is defined as the frequency with which two items are cited together in subsequent literature. Co-citation is used to identify a core of earlier literature and to reveal relationships between key ideas. According to Henry Small and Belver Griffith (1974) it may lead to a more objective method than is currently in use of mapping the structure of scientific specialties and of monitoring developments and interrelationships between fields.

Catalogs are bibliographies organized so as to aid in locating documents within a particular collection—for example, within a library. Although a variety of cataloguing schemes are in use, the Dewey Decimal and Library of Congress systems are the most commonly used.

"Union lists are catalogs which provide a directory for the titles to be found in the libraries of a particular area and show the libraries that hold any given title . . . *The National Union Catalog*, A Circulation Author list published by the Library of Congress . . . includes all works currently catalogued by the library and by libraries participating in the Shared Cataloguing Program" (SATCOM, 1969:136). Each day the Library of

Congress distributes over 500 million bits of information in the form of catalog cards to over 20,000 subscribers. *The Union List of Serials in the U.S. and Canada* is the most comprehensive catalog of periodicals covering more than 156,000 titles in 956 cooperating libraries. "Present estimates indicate that there are between 400,000 and 500,000 serial titles in existence today, most of which are for inactive serials" (SATCOM, 1969: 136).

Reprocessing is "the selection of information especially relevant to the needs of a particular group of users (or an individual) and delivery of this information to them (him)" (SATCOM, 1969:177). Reprocessing involves reworking of primary messages and, in many cases, secondary messages as well. It incorporates the concept of Selective Dissemination of Information (SDI), which is generally accomplished by matching messages with users' interests (profiles) to produce individualized listings of relevant primary messages. This function was once performed by librarians, but with the rapid expansion of the literature it became difficult to provide such services manually. The advent of computerized information storage, retrieval, and dissemination has given rise to a new industry that provides these services. This industry is described later in this section.

Major data bases provide SDI services. For example, CLASS (Current Literature Altering Search Service) of BIOSIS, ISI's ASCA IV, NASA/ SCAN (Selected Current Aerospace Notice), ERSO/ECDO (Space Document Service), and CAN/SDI (Canadian Selective Dissemination of Information) of the National Science Library of the National Research Council in Ottawa are established SDI services. Frequently these respond to "canned" or group profiles.

Abstracts are condensed summaries of the substance of primary messages. They may appear alone or in combination with bibliographies, in annotated bibliographies. Abstracts are generally one of three types: (1) indicative (announcement), (2) informative, or (3) critical.

The number of abstracts produced annually has been growing rapidly. Between 1963 and 1973 it increased 76 percent, and in 1974 more than 1.4 million abstracts were expected (Olson and Shank, 1972). Increasing use of abstracts suggests that they are often used in place of the primary message.

There is an increasing trend toward author-prepared abstracts:

In regard to the preparation of abstracts and indexes, there is a trend toward increased sharing of responsibility between authors and secondary services. The requirement that reports be accompanied by abstracts is spreading among government agencies, and regulations and guides for this purpose have been published. A 1962 study of scientific and technical journals showed that most of the abstracts that accompanied primary publication were author-prepared, and another study produced evidence that secondary publications make extensive use of such abstracts. By 1961, 60 percent of the papers cited in *Physics Abstracts* were based upon author abstracts. Further, recent estimates suggest that about half the 50,000 abstracts currently carried by the *Nuclear Science Abstracts* are author prepared. Proponents of the author-prepared abstract believe that an author is best able to digest the essence of his paper; critics claim that an author is apt to put into the abstract what he wished the paper contained (SATCOM, 1969:140-141).

The Institute of Electrical and Electronic Engineers produces *Science Abstracts*, which is composed of *Physics Abstracts, Computer Controls Abstracts,* and *Electrical Engineering Abstracts.* Chemical Abstracts Service, BioSciences Information Service, American Psychological Association, and American Society for Metals are major nonprofit organizations providing abstracting services.

Most abstracts are of the informative type; they describe the content of a primary message. Some, however, evaluate the content. Those that do are called "critical abstracts." Such abstracts are generally longer than ones that are informative. When evaluation begins to dominate description the message becomes a *review.* Reviews, however, may cover a number of publications usually related by subject matter and therefore can cover a subject or field.

The principal outlets for review articles are journals containing primary messages in which there is also a review section. Virgo (1971) estimated that two-thirds of the review literature is found in primary journals. There are, however, a few journals that contain only reviews or reviews and abstracts—for example, *Computing Reviews.* In most fields only a small percentage of all reviews are included in annual review publications and books. Although lengths vary greatly, the average length of a review covering a subject is about twenty pages.

Citation analyses reveal that the review generally remains useful over a long period of time. It is a valuable means of coping with the increasing amount of information (SATCOM, 1969).

Synthesis or *Consolidation* is the process of "evaluation, compaction, simplification, and the fitting of isolated items into a general framework" (SATCOM, 1969:177). The products of such syntheses are tertiary messages that take a variety of forms—for example, critical reviews, state-of-the-art surveys, encyclopedias, handbooks, and textbooks. The more a review reworks primary messages, the more primary it becomes.

Systematic and regular production of state-of-the-art papers is conspicuously lacking in most fields. Many scientific and technical societies include at least a certain number of state-of-the-art reports or discussions in their meetings. In nine scientific and engineering disciplines the percentage of program material devoted to state-of-the-art and review presentations ranged from 25 percent to over 50 percent. It was heaviest in two engineering disciplines. The amount of state-of-the-art literature varies greatly by field. In general, the amount falls short of the need (SATCOM, 1969).

Although there is little to say about informal referral (e.g., by friends and colleagues) that is not obvious, it should be remembered that it is both extensive and effective. Personal communication is one (if not the) principal means of transmitting information about the products of science and technology.

Information Services

As the output of science and technology has grown, so has the need for secondary and tertiary services, tools, and documents. It is not surprising, therefore, that the number and variety of organized "secondary" information services, particularly ones that are computer-based, are increasing rapidly.

There were approximately 112 information and data analysis centers in the United States in 1968 of which only 21 were established prior to 1958 (SATCOM, 1969). According to Stella Kennan of the National Federation of Abstracting and Indexing Services (NFAIS), there are approximately 150 abstracting and indexing services in the United States today. Most of these services are rather small and are project- or mission-oriented. However, the number is somewhat understated, since the basis for selection excludes many small service organizations. NFAIS includes twenty-nine organizations, which produce over half of the abstracting and indexing services in the United States (see Table 7-6). These services have grown rapidly over the past ten years. Furthermore, according to the Association of Scientific Information Dissemination Centers, within the past five or six years approximately 30 new information-dissemination centers have appeared (Park, 1973).

Abstracting and indexing organizations can be divided into five broad classes:

(1) industrial (specialized within a corporate structure),
(2) academic,
(3) profit,
(4) nonprofit, and
(5) governmental.

Organizations within each class provide a variety of services designed to accommodate their users. Some provide discipline-oriented services (e.g., Biological Abstracts and Chemical Abstracts) while others provide mission-oriented services (e.g., American Petroleum Institute). There are other significant differences between the classes. We consider each of them in turn.

Industrial. Secondary-service units in industry provide a variety of services including technical libraries and report centers, patent depart-

Table 7-6.

**National Federation of Science Abstracting and Indexing
Member Services, 1976**

Voting Members	*United States Government Affiliates*
American Dental Association	Defense Documentation Center
American Geographical Society	Energy Research and Development
American Geological Institute	Administration
American Institute of Physics	Library of Congress
American Meteorological Society	National Aeronautics and Space
American Petroleum Institute	Administration
American Psychological Association	National Agricultural Library
American Society for	National Institute of Education/DHEW
Information Science	National Library of Medicine
American Society for Metals	National Oceanic and Atmospheric
BioSciences Information Service of	Administration
Biological Abstracts, Inc.	National Technical Information Service
Center for Applied Linguistics	Water Resources Scientific Information
Chemical Abstracts Service	Center
Documentation Abstracts, Inc.	
Engineering Index, Inc.	
Exxon Research and Engineering	
Company	*Foreign Affiliates*
The Foundation Center	
Index to Religious Periodical	Centre National de la Recherche
Literature	Scientifique (CNRS)
The Institute of Electrical and	INSPEC
Electronics Engineers, Inc.	International Labour Office (ILO)
Medical Documentation Service	Commonwealth Scientific and Industrial
National Association of Social	Research (CSIRO)
Workers, Inc.	National Library of Australia
Penn-State Coal Carbon Data Bases,	United Kingdom Chemical Information
Pennsylvania State University	Service
Philosophy Documentation Center	
Primate Information Center	
University of Tulsa	

ments, selective dissemination of information (SDI), data-bank operations, and computerized information services. Although such units are
primarily concerned with in-house needs, some also provide services to
outsiders, particularly their customers (SATCOM, 1969).

"The Celanese Research Center has built an information center which
brings together the library and the computerized and micrographic infor-
mation-handling system" (SATCOM, 1969:196). IBM has developed the
Technical Information Retrieval Center (ITRIC). ITRIC is a corporate
literature-retrieval system that utilizes five external databases as well as

internal sources to provide dissemination and current-awareness services. General Telephone and Electronics Laboratories developed a similar service that links internal report collection with external databases to provide a corporate technical reports processing system including a report ordering and distribution system (Gechman, 1972).

E. R. Squibb and Sons provides an example of the problems encountered and the types of solutions developed by a private corporation in the development of information services. Squibb is a pharmaceutical company that engages in extensive research and development. One of the problems it faces is how to maintain adequate coverage of the relevant literature. This is an extremely difficult and costly task even for the subset of pharmaceuticals relevant to the company. Therefore, Squibb subscribes to the RINGDOC service of Derwent Publications. This database provides coverage (indexes and abstracts) of about 350 relevant journals. The company uses the INQUIRE software package provided by Infodata Systems coupled with the Automatic Subject Citation Alert tape service of the Institute for Scientific Information (current-awareness service) and internally generated databases. The composite system provides more than 100 individual users with dissemination services (current awareness) and retrospective searches within a database that is growing by 45,000 documents per year (Frycki, 1973).

Academic. The academic community is a major provider and user of information services. To facilitate access to the vast scientific literature, many universities have developed computerized information retrieval systems. Because of a lack of resources, they generally use externally developed databases rather than develop their own. Some academic information services have obtained substantial funding from external sources (usually government) and frequently help defray costs by selling their services outside the university community. For example, the Library Group of Southwestern Connecticut, Inc. serves the needs of a group of corporations in Connecticut. It uses the resources of ten major Connecticut libraries. A similar project at Rice University is called the "Regional Communications and Information Exchange" (SATCOM, 1969:200).

The National Science Foundation has supported a number of universities in the development of services that utilize machine-readable databases. These include the Universities of Georgia and Pittsburgh, Ohio State University, UCLA, Lehigh University and Stanford University. The University of Georgia Computer Center currently provides search services that utilize seventeen databases. These external databases have to be converted to a standard format before being used (Gechman, 1972:327-328). The center also provides selective dissemination services to users.

Ohio State University's Mechanized Information Center has developed an automated, multidisciplinary, university-centered information system. Using five databases, it provides selective dissemination services for the university community and other colleges in Ohio. Other academic information services include those of Illinois Institute of Technology Research

Institute (IITRI) and the University of Florida. Furthermore, there are several universities that have developed databases in specific disciplines. These include Bowling Green State University (philosophy) and University of Tulsa (petroleum industry). These are by no means all the universities that provide services utilizing specialized databases (Gechman, 1972).

Profit. Profit-oriented information service organizations are a part of what is called the "information industry." They are concerned with providing information services for external use. Such organizations are faced with competition not only from each other, but also from governmental and other nonprofit organizations. Nevertheless, the "information industry" is generally believed to be one of the next big growth industries.

Organizations in this category provide two major kinds of service. One is the production of machine-readable databases and the other is the utilization of these and other databases as in selective dissemination service and retrospective search.

The Institute for Scientific Information (ISI) is a major profit-oriented producer of multidisciplinary databases. ISI developed and offers citation indexing. The major advantage of citation indexing (according to Morton Malin of ISI) is that it allows for automatic classification and indexing, thus eliminating the need for human judgement and intervention and can also be used for retrieval. With one exception, ISI carries no abstracts in its databases. This enables it to keep its files current, since it does not have to wait for abstracts to be prepared or to enter them once they are available. The principal services provided by ISI are:

(1) *Current Contents* (described above), a portion of which is computerized, covers 5,000 journals.

(2) *Science Citation Index* has 500,000 items per year and covers 2,600 journals.

(3) *Social Science Citation Index* has 90,000 items per year from 1,100 journals.

(4) *Automatic Subject Citation Alert* (ASCA-IV), provides selective weekly dissemination of information, covering 3,200 journals.

(5) *ASCATOPICS* is the same as ASCA-IV except that it utilizes 470 standard interest profiles.

(6) *Current Abstracts of Chemistry* and *Index Chemicus* weekly abstract service for synthetic organic chemistry has a one-month lag time and covers 150,000 newly synthesized compounds per year.

(7) *ISI Search Service* provides custom-designed retrospective searches.

(8) *Original Article Tear Sheet* (OATS) provides original articles from any journal covered by ISI.

ISI also provides several supplementary services.

In 1974 ISI processed 450,000 source articles and about 4 million citations into its database. The cumulative database at that time included close to 3 million source items and 30 million references.

There are other organizations that generate databases and provide search services. American Bibliographic Center, Clio Press, provides coverage of political science and history. Compendium Publishers International offers Search-Data, which covers 1,000 trade journals in the chemical industry. Aspen Systems Corporation produces and maintains a database covering the laws of all fifty states (200 million words of statute law in full text form). Utilizing this database it provides searches that encompass one or more states or special combinations. It also provides a current-awareness service covering recently enacted legislation. As mentioned earlier, the RINGDOC Distribution System developed by Pfizer Pharmaceuticals is offered by Derwent Publications, Ltd. It provides three databases (RINGDOC, VETDOC, and PESTDOC) to seven subscribers, all pharmaceutical firms, but no information services.

Systems Development Corporation (SDC) provides a nationwide information service using a centrally located multi-databased service. SDC converts databases (from ERIC, *Chemical Abstracts Condensates,* the CAIN agricultural database, and MEDLINE) so that they can be searched by means of SDC's ORBIT #II on-line information retrieval system. A subscriber can obtain on-line access to one or more of these databases from a single terminal. This is only one of a growing number of similar services whose continued growth is to be expected (Vinsonhaler and Moon, 1973). The other major on-line commercial information retrieval system is Lockheed's DIALOG. It is similar to SDC's ORBIT system.

Nonprofit. This class includes information services provided by professional societies and nonprofit associations and foundations. Most such services cover a specific discipline. Some, however, are mission oriented. In general, their coverage ranks high in quality and usefulness.

The Chemical Abstract Service (CAS) of the American Chemical Society is the world's largest secondary information service in the sciences with sales of approximately $20,000,000. It processed over 400,000 abstracts in 1971 and is projecting 600,000 abstracts a year with a total of 10 million by 1980. To cope with such growth, according to Graham et al. (1967), CAS has been working toward total computerization of its information-handling and publishing operations. It is supported in this effort by NSF.

CAS's output consists primarily of identifications, abstracts, and index entries for items selected from the world's chemical literature (most of which come from about 3,000 journals). Depending upon the particular service, however, anywhere from 49 to 12,000 journals and patents are covered (Gechman, 1972). In addition to the printed *Chemical Abstracts* (CA) and its indexes, which contain virtually all of the CAS database, a number of specialized publications and services are produced in printed form and on magnetic tapes (NFAIS, 1974). These include:

(1) *Chemical Abstracts*—weekly issues,
(2) *Chemical Abstracts*—volume indexes (semiannual),
(3) *Chemical Abstracts*—on microfilm,
(4) *Chemical Abstracts Condensates*—weekly tape carrying approximately 8,000 abstracts per week,
(5) *Basic Journal Abstracts*—covering abstracts from 49 basic journals, and
(6) *Chemical Titles*—biweekly service carrying 5,000 titles from 700 journals.

This list does not exhaust the types of output provided by CAS. Chemical Abstracts leases or licenses its database to information centers that provide services from their tapes. Users can search the database through these centers. The CA database can be combined with other databases. Most of the services that are dependent on CAS have been previously mentioned (IITRI, GTE Laboratories, MIC at Ohio State University, University of Georgia, SDC's ORBIT #II and Lockheed's DIALOG). These centers are independent of CAS but are licensed to use its computer-readable files (CAS-1974 Information Services).

The Biosciences Information Service of Biological Abstracts (BIOSIS) is the world's leading abstracting and indexing service in the life sciences. In 1969, BIOSIS published its two millionth abstract. Unlike CAS, which is a division of the American Chemical Society, BIOSIS is an independent organization that maintains affiliations with professional societies through its Board of Trustees.

The services provided by BIOSIS are much like those of CAS. It has five major publications.

(1) *Biological Abstracts*—140,000 abstracts annually, 4 indexes are provided with each issue (BASIC, a subject index; KWIC, CROSS, another subject index using over 623 major subject headings and subheadings; Biosystematic Index, a taxonomic index; and Author Index).
(2) *Biosearch Index*—one volume per year with 100,000 reports, theses, and so on that do not appear in *Biological Abstracts*.
(3) *Cumulative Indexes*—an annual index.
(4) *Lists of Serials*—an annual whose 1972 issue contained 7,661 active titles from 101 countries.
(5) *Biosearch Today*—a series of monthly journals covering specific topics.

In addition, it publishes more specialized collections such as *Abstracts of Micology* and *Abstracts of Entymology*.

BIOSIS also provides a magnetic tape service called *Biological Abstracts Previews*, which covers the material in *Biological Abstracts* and

Bioresearch Index. It is issued one month in advance of the published copies. In addition BIOSIS provides computerized search services called CLASS (Current Literature Alerting Search Service), Standard Profile Service, and Retrospective Search Service. The Retrospective Search Service and CLASS are tailored to an individual's need, whereas the Standard Profile Service is designed as a group service (NFAIS, 1974).

Other major nonprofit information services are provided by the American Institute of Physics; by the Engineering Society, which produces *Engineering Index* and *Compendex*, a tape of *Engineering Index Monthly* (containing 3,500 abstracts per month from 7,500 sources of literature); and many more (Gechman, 1972).

Government. The federal government is probably the largest single producer and user of information services. In his review of machine-readable databases Marvin Gechman (1972:334) wrote: "U.S. Government agencies produce more databases than any other group referred to in this review. Most of them appear to be designed for specific internal uses and serve a relatively small audience. However, most of the largest machine-readable databases are also produced by the government."

According to Gechman (1972), three national libraries make significant contributions to the dissemination process.

(1) The Library of Congress provides machine-readable catalog (MARC) tapes that contain the Library's cataloging information for all English language monographs and related indexes. Initiation of these tapes was probably the singly most important step toward the computerization of library search services.

(2) The National Agriculture Library (NAL) is responsible for worldwide collection and dissemination of agricultural information. NAL's cataloging and indexing (CAIN) tapes are distributed to other services. The database is compiled from the NAL's "Bibliography of Agriculture" and "Food and Nutrition Catalog."

(3) The National Library of Medicine (NLM) collects, indexes, and disseminates published medical information to health-science professionals throughout the world. This library has an exhaustive collection covering about forty biomedical areas and to a lesser degree covers such related subjects as chemistry, physics, and botany.

The primary publication of NLM is *Index Medicus.* Its computerized database, called MEDLARS (Medical Literature Analysis and Retrieval System), contains over 1.5 million citations of articles published since 1963. Approximately 200,000 citations are added annually from about 2,300 periodicals (half of which are non-English) (Gechman, 1972). In 1971 NLM initiated MEDLINE (MEDLARS on-line). MEDLINE provides access to 350 terminals around the country for on-line search (in an interactive mode) of the MEDLARS database. It also provides less expensive off-line searches. To augment this search capability, the NLM has developed a system of eleven regional libraries to facilitate user access to documents. Thus, the NLM attempts to bridge the gap between dissemination and acquisition.

Information services in the field of education are provided by the ERIC system. Project ERIC (Education Research Information Center) is sponsored by the United States Office of Education. It provides grade-school to university educators with access to material that can help them stay abreast of their field and develop educational programs. It is a network composed of a central facility, eighteen decentralized clearinghouses, and a document reproduction service. The clearinghouses, each of which specializes in an educational field or topic, are responsible for collecting the literature in their fields, ERIC provides numerous services such as interpretative summaries, research reviews, bibliographies, other reference tools, and supplies primary documents and/or microfiche (Vinsonhaler and Moon, 1973).

There are many other government information services. The following is a brief list provided by Gechman (1972:336-339) of other government agencies that provide such services:

(1) Defense Documentation Center,
(2) Energy Resources Development Administration—Nuclear Science Abstracts,
(3) National Technical Information Service,
(4) NASA—6 Regional Dissemination Centers—and the on-line system NASA/RECON,
(5) Smithsonian Science Information Exchange,
(6) Patent Office,
(7) National Institute of Neurological Disease and Stroke (NINDS), and
(8) Environmental Protection Agency (EPA-APTIC).

Overview of Dissemination

In this section we have provided an overview of the current process of dissemination. We began by examining distribution and referral processes and then described how information service organizations are involved in these processes. In the literature dealing with these services several recurring themes stand out.

First, the amount of literature to be covered is growing at an increasing rate (about 7 to 7½ percent per year). The difficulties involved with providing comprehensive coverage of the literature are tremendous. Although the computer has had a significant impact on these difficulties, without organizational and other types of change it is not likely to be enough to remove them.

A number of information services are available, but these services are not having the impact they could have because of overlapping coverage and lack of standardization. The reduction of duplication and the introduction of standards are vital to efficient and effective functioning of the SCATT System. Coordination of these services is badly needed.

The number of machine-readable databases is large and increasing. Like the services based on them, they are not under any coordinating influence. It is important that standards be set for bibliographic structure and record format for magnetic tape. According to NFAIS (1973:19): "Such standards will provide for the interconnection of systems, so that, by the end of this decade the major discipline services will be truly coordinated in coverage, input procedure, and intellectual content." Furthermore, the centers that utilize multiple databases are paying more than they should for duplicated items as well as incurring excessive costs in file conversion.

Finally, there is the need to provide access to documents. Only a few information service organizations (ISI and NLM) provide documents or make their location known. Many SDI services provide photocopies of documents but this practice is currently under legal attack. Now let's consider the acquisition of documents in more detail.

ACQUISITION

Many, if not most, documents that scientists and technologists receive are unsolicited. But here we focus on user-initiated (solicited) acquisition.

There are four principal types of source from which documents are acquired: (1) friends, colleagues, and authors; (2) publishers (by subscription or purchase); (3) libraries; and (4) information services.

Friends and colleagues are sources of many documents, some individually requested, others not. The latter are often sent in response to a tacit or explicit request that is general in nature—for example, "Send me whatever you think I ought to see." Friends and colleagues are among the most effective filters a scientist or technologist can have. Furthermore, academic scientists and technologists are provided with students who perform literature searches for them and often serve as their filters.

Both scientists and technologists acquire documents in the form of preprints or reprints directly from the author, some by request, others not.

Most scientists and technologists subscribe to three or more journals published by professional societies or others. Not all of the material acquired in this way is read, but most subscribers examine the tables of contents and read or scan at least some of the content and references. This is less true for journals provided without subscription.

Many subscriptions are not optional but come with membership in professional societies. Such "imposed" subscriptions fill personal libraries with unread journals and give their publishers exaggerated opinions of the usefulness and importance of their product.

If subscriptions to society-related journals were made optional, the number of subscribers would be likely to decline considerably. The Society for Automotive Engineers (SAE) did separate membership and journal subscription, and only 5,700 out of a total membership of 25,000 bought its *Transactions*. Of the 5,700, more were over 65 than were under 35. There appears to be an "inertial" support for the old SAE journal (David Staiger, personal communication, September 27, 1974).

Scientists and technologists also make purchases of single documents. More of these are likely to be read than are those received without request or by subscription because they are usually acquired on recommendation or to serve particular purposes.

The use of libraries has been studied extensively. For example, a study of Bath University's Library Groups in England points out that there is a positive correlation between the age of the researcher and his estimation of the library's adequacy, and a negative correlation between his age and use of it. This suggests that the older and more experienced researcher is more likely to call upon his colleagues and friends and to have access to other sources of documents. The Bath study also revealed that those who seek methodological and conceptual information were more easily satisfied by libraries than those who seek historical and descriptive material (Lin and Garvey, 1972). Library use has also been found to correlate positively with education (Lipetz, 1970).

Libraries are used primarily by persons living or working near them. In a study of public reference libraries, it was found that half of the users traveled less than two miles, and that 30 percent of those who use the library did not use the literature it contained. Half of the users were students, many of whom use the library as a place to work. On the average, four documents were consulted per visit (Wood, 1971). Most users look for known items or publications. Assistance of the library staff is requested more frequently in industrial and professional society libraries than in academic libraries (Lipetz, 1970).

As previously noted, some information services provide users with portions of documents. For example, the Institute of Science Information provides original articles to anyone. This service is known as OATS: Original Article Tear Sheet. ISI provides either an original article or a photocopy of it; arrangements have been made with the publishers to pay copyright royalties. There are, however, relatively few services that provide the complete message because of copyright laws.

Acquisition Studies

Wood (1969) classified the ways of obtaining information about users' acquisitions:

(1) questionnaires,
(2) interviews,
(3) diaries kept by users,
(4) study of records, and
(5) direct observation.

Most studies of acquisition can be placed into one of four categories:

(1) one-discipline one-channel,
(2) one-discipline many channels,
(3) many-disciplines one-channel, and
(4) many-disciplines many-channels.

Relevant studies in each of these classes are discussed below.

One-Discipline One-Channel. Studies of the use of one channel of communication by users in one discipline are limited in scope, but when compared can yield some general information.

The MIT Science Library studied the use of its physics journals (Chen, 1972) and found that 62.7 percent were used once or more during a three-month interval; 49 journals covered 90 percent of the usage but only 51.5 percent of the total subscription cost of those which were used. This and many other studies show that a core of journals and other publications are used most heavily to supply information. Wood (1971:18) reported similar findings:

In the medical field Hardegg found that only 10 journals were used frequently by more than 50% of the teaching staff at Heidelberg University. In the field of astronomy 75% of the demand can be satisfied by only 9 titles, and Strain reported that only 25% of the titles in the IBM Electronics Library were heavily used. At the NLL (National Lending Library for Science and Technology in England) a similar state of affairs exists. In a survey of the use of medical literature, 3.3% of the titles used were found to account for 21.5% of the loans and in the social sciences, 17 titles accounted for 20.2% of the request, 116 for 55%. Trueswell refers to this characteristic as the 80/20 rule, i.e., 80% of the demand can be satisfied by 20% of the stock and points out the fact that the same situation pertains in the field of business and industrial inventory.

In two studies conducted at the Case Institute of Technology in 1958 and 1960, it was found that a core of ten journals accounted for 55.1 and 49.8 percent (respectively) of the chemists' and physicists' journal reading time (Tables 7-7 and 7-8).

Table 7-7.

Ranking of Journals by Chemists' Reading Time
(Martin and Ackoff, 1963, p. 334)

1960 Study % of Reading Time	Rank	Journal	Rank	1958 Study % of Reading Time
15.7	1	*Chem. and Engr. News*	2	10.5
7.0	2	*J.A.C.A.*	1	12.4
4.3	3	*Chemical Engineering*		
3.7	4	*Analytical Chemistry*	7	3.5
3.3	5	*Chemical Abstracts*	3	8.6
3.3	5	*J. Chem. Society (London)*	8	2.8
3.3	5	*Chem. Processing*		
3.0	8	*Ind. and Engr. Chem.*	9	2.6
3.0	8	*J. of Polymer Science*	6	3.7
2.7	10	*Chem. Week*	4	4.9
		Journal of Organic Chem.	5	3.7
		Oil and Gas Journal	10	2.4

Source: Miles W. Martin, Jr. and Russell L. Ackoff, "The Dissemination and Use of Recorded Scientific Information," *Management Science* 9 (1963):334. Reproduced by permission of The Institute for Management Science.

Table 7-8.

Ranking of Journals by Physicists' Reading Time
(Martin and Ackoff, 1963, p. 334)

Rank	Title	% of Physicists' Sci. Reading Time	Cumulative %
1	*Physical Review*	11.9	11.9
2	*J.A.C.A.*	7.8	19.7
3	*Review of Sci. Instr.*	6.6	26.3
4	*Physics Today*	6.1	32.4
5	*J. Phys. & Chem. of Solids*	4.1	36.5
6	*Physical Review Letters*	3.8	40.3
7	*Proc. of Inst. of Radio Eng.*	3.5	43.8
8	*Chem. and Engr. News*	3.0	46.3
9	*Scientific American*	3.0	49.8

Source: Miles W. Martin, Jr. and Russell L. Ackoff, "The Dissemination and Use of Recorded Scientific Information," *Management Science* 9 (1963):334. Reproduced by permission of The Institute for Management Science.

The core literature concept is consistent with Bradford's law, which states that the most productive articles on a given subject are concentrated in a "nucleus" of periodicals and that there is a common pattern to the distribution of these useful papers in the periodicals. The Bradford distribution identifies the number of journals needed to cover a certain proportion of these useful papers (Leimkuhler, 1967).

One-Discipline Many-Channels. Studies of the various channels used by scientists in one discipline are relatively common. (Channels may refer to persons, institutions, publications, and so on.) For example, Menzel (1970) studied the sources used by polymer chemists in their searches. Table 7-9 displays his results. Selection of the "main final channel," the channel that provided most of the information used, varied depending on the type of search conducted. Informal channels were the primary focus of procedural searches. Searches for theory relied on both formal and informal channels.

The low usage of secondary sources (e.g., bibliographies and indexes) for "lead channels" is revealed in Table 7-10. Menzel (1970:9) defined "lead channels" as those "which played a role in the process of leading him (the scientist) directly or indirectly to the channel which ultimately delivered the message." Wood (1971) in his review of user studies noted

Table 7-9.

Main Final Channels in Deliberate Searches by Polymer Chemists for Procedures, Findings, and Theory

	For Procedure[1]	For Findings[2]	For Theory[3]
Published Primary Literature[4]	28%	64%	51%
Semi-Published Primary Literature[5]	8	1	4
Retrieval and Library Services[6]	1	4	0
	37%	78%	55%
Personal Communication	63%	23%	43%
Presentations (both small and large meetings)	1%	0%	2%

[1] "Information about procedures, techniques, materials, and apparatus" (Menzel, 1970, p. 4).
[2] "Information about facts, data, finds, or constants" (Menzel, 1970, p. 4).
[3] "Information of a theoretical nature—concerning mechanisms, processes, explanatory schemes, or the like" (Menzel, 1970, p. 4).
[4] Includes articles and reprints, handbooks, other books, literature (unspecified).
[5] Includes suppliers' literature, own organization's technical reports or patents, or other technical reports.
[6] Includes published retrieval tools (published abstracts, bibliographies, annual reports, proceedings, publisher's notices, patent abstracts and indexes and review books and articles) and library facility or service.

Source: Menzel, 1970, pp. 25-26. Copyright © 1970 by Herbert Menzel.

Table 7-10.

**Lead Channels in a Deliberate Search by Polymer
Chemists for Procedures, Findings, and Theory**
(Menzel, 1970, p. 31)

	For Procedure	*For Findings*	*For Theory*
Published Primary Literature	26%	60%	52%
Semi-Published Literature	11	11	5
Retrieval and Library			
Services	2	4	2
	39%	75%	57%
Personal Communication	60%	24%	39%
Presentation (both small and			
large meetings)	1%	0%	3%

Source: Menzel, 1970, p. 31. Copyright © 1970 by Herbert Menzel.

that many library surveys have also found low usage of secondary services, and he suggested that lack of familiarity with them was the principal reason.

An equally plausible explanation is that users of secondary sources find it difficult to identify what is relevant and useful from these sources alone. Secondary sources may lead to examination of too much irrelevant or useless information.

The INFROSS (Investigation into Information Requirements of the Social Sciences) Project of the Bath University Library Group focused on the needs and uses of channels by social scientists. It found that bibliographies and references in books and journals were used more heavily than any other tools by social scientists. Major deficiencies in their formal information channels are "the lack of review articles, translation services, and services oriented to the practitioner" (Lin and Garvey, 1972:11). Some of the results of the Bath study are shown in Tables 7-11 and 7-12.

In 1961 the American Psychological Association began a large-scale study of scientific information exchange in psychology. "The basic approach taken by this project has been to make an intensive and comprehensive study of the scientist's information-exchange behavior within his scientific environment" (APA, 1965:1). The project focuses on use of many channels, such as conventions (Reports 3, 4, and 5), journals (Report 9), technical reports (Report 13), and books (Report 14); and on information exchange activities and the usage of different channels and the characteristics of users (Reports 17, 18, 19 and 21).

The findings in Report 19, "Information Exchange Activities involved

Table 7-11.

Use of Different Information Channels by Social Scientists

Channel	Not Using (%)	Using Often (%)
Periodicals	3	57
Books (monographs)	3	57
Books (conference procs.)	13	32
Research reports	21	28
Theses	29	11
Newspapers	36	11
Government publications	23	34
Microforms	74	3
Maps	63	10
Films (pictorial)	93	1
Other pictorial	76	2
Sound recordings	85	4
Video-tape	97	1
Computer printout	62	21
Radio or TV	78	1
Conferences	41	8
Colleagues in own institution	15	26
Colleagues or experts elsewhere	16	26

Source: Maurice B. Line, "Information Uses and Needs of Social Scientists: An Overview of INFROSS," *Aslib Proceedings* 23 (1971):412-434. Reproduced by permission of Aslib.

Table 7-12.

Usefulness of Methods of Locating References for
Primary Research by Social Scientists

METHOD	USEFULNESS % of respondents	
	Not Using	Very Useful
Abstracts or indexes	22	32
Consulting expert	16	34
Discussion with colleagues internally	13	27
Discussion with persons elsewhere	17	23
Library catalogs	22	22
Searching own library shelves	23	18
Searching other library shelves	34	16
Consulting librarian	48	8
Specialist bibliographies	35	23
Bibliographies or references in book/journal	6	59
Book reviews	23	17

Source: Maurice B. Line, "Information Uses and Needs of Social Scientists: An Overview of INFROSS," *Aslib Proceedings* 23 (1971):412-434. Reproduced by permission of Aslib.

in Psychological Work" (APA, 1969), correspond with Menzel's. Of those researchers holding research titles in universities and colleges, "very important" ratings were given by 84 percent to journals, 43 percent to books, 37 percent to discussions with local colleagues, and 19 percent to discussions outside their institutions. Researchers in private industry or industrial consulting firms, however, placed more emphasis on informal media and less on journals and books.

Many-Disciplines One-Channel. Studies of the use of one type of channel by many disciplines are fairly common. For this type of study, records of use may be examined or questionnaires sent to a sample of users. For example, D. T. Hawkins (1974) recently conducted a study of bibliographic databases used to satisfy requests for information from chemists, physicists, material scientists, electrical engineers, educators, and telecommunication scientists at Bell Telephone Laboratories. He found that four major sources satisfied half of the requests and that half of the searches required the use of only one database. This result is not so surprising as it would be if the different disciplines were not all involved in related research.

The use of periodicals by graduate students at the University of Michigan was studied by Stephen Peterson. Half of the responding students (85 percent of 400) knew the precise reference they wanted before going to the library. Of the remaining half, 65 percent identified the reference through abstracts and indexes at the library. Rarely did they read more than one or two papers per issue or volume consulted. Only about 15 percent of the respondents indicated they scanned periodicals which they had chosen by subject interest (Lipetz, 1970).

In a study of the circulation of about 9000 books over a six-month period at the Research Library of the Air Force Cambridge Research Laboratory, it was found that half of the books were borrowed by only 12.5 percent of the borrowers and that 90 percent were borrowed by about half of the borrowers. Many other circulation studies have been conducted, and many were reviewed by Wallace Olsen (Lipetz, 1970).

Many-Disciplines Many-Channels. Studies of the use of more than one channel by more than one discipline have been conducted by organizations containing different types of users. They either focus on channel use by all users, or they compare channel use by different disciplines. For example in the Department of Defense user study, the Auerbach Corporation (1965) examined the use of the formal and informal systems by the department's scientists. It was found that systems were not used fully because of a lack of awareness on the part of the scientists of the sources and because the sources do not provide such features as "convenience, responsiveness and ability to conduct dialogue with the system."

Rosenbloom and Wolek (1970) found that the source employed is related to the purpose of the user. Work with a "professional focus" tends to draw upon external sources, including journals and external personal communications, whereas work with an "operational focus" tends to draw upon internal corporate sources.

Allen and Gerstberger (1968) identified three criteria used by R&D engineers in the selection of an information source.

(1)   Accessibility is the single most important determinant of the overall extent to which an information channel is used.

(2)   Both accessibility and perceived technical quality influence the choice of the first source.

(3)   Experience influences perceptions of accessibility. The more experience an engineer has with a channel, the more accessible he perceives it to be.

Utterbach also found accessibility to be critical: "In sum, search for information is more highly ordered following problem identification than before, and search proceeds in a hierarchical order from more to less easily accessible information sources" (cited in Allen, 1969:6). The implication of Allen's and Utterbach's studies is that an improvement of the quality of messages provided by a particular information source may not, by itself, lead to increased use of it; it must also be easy to use.

The Auerbach Corporation (1965) found heavy reliance on the closest sources for information. In more than half of the searches, colleagues, personal files, and local departmental sources were the first used. Informal sources satisfied needs in 39 percent of the searches. Oral communication was very important and accounted for 30 percent of the acquired information.

In order to understand the differences in channel use among scientists and technologists, according to Lin and Garvey (1972:24) it is necessary "to understand the internal structure of and the interrelation of the various channels as they provide a network of information sources for scientists and technologists as individuals and groups."

We are a long way from such understanding. We suffer from a lack of both analyses and syntheses of the results of "needs and uses" studies. Explanatory hypotheses should be formulated and tested. Where discrepancies in results exist they should be explained; they may be due to the study's sample, disturbances of the environment caused by the survey, different interpretations of questions, differences in the analyses of data, or differences between disciplines (Barnes, 1965). Until we go beyond purely descriptive studies and attempt explanation, we will continue to float aimlessly on a sea of data.

Summary

Information is acquired from a variety of formal and informal sources such as friends and colleagues, authors, publishers, libraries, and information services. Many studies have been conducted on scientists and technologists to determine which sources are used. These studies are frequently discussed under the title of "Needs and Uses," but in general they are primarily concerned with means of acquiring information, not its suitability or use.

A few apparently significant findings have come out of acquisition studies.

(1) A relatively small number of journals and other publications is used to satisfy most requests for information. This raises questions about the growing number of journals: how many if any are necessary? Which ones are useful and what are their characteristics? In all likelihood the number of journals in the core will not increase as the total number does, but the number of cores seems to be increasing with the emergence of new disciplines and inter-disciplines.

(2) Formal acquisition systems are not used as heavily as informal ones unless they have the same desirable characteristics as the informal: ease of use, accessibility, and responsiveness.

(3) Channel use varies with the type of search being conducted and the place of research. The basic researcher looking for theoretical information and conceptual guidance is likely to use formal channels. Researchers in universities consider journals to be very important, while those in private firms rely more heavily on informal channels.

(4) Researchers have a variety of information needs that change throughout their work. The APA (Report 11) found that formal channels are used most heavily by researchers during the idea-generating stage of research while informal channels are useful in amplifying and clarifying information and defining problems.

These results and others have not been synthesized into a sufficiently detailed description (let alone explanation) of information acquisition to enable one to determine how to improve this process significantly. Most descriptive studies appear to have been carried out for their own sake rather than for the sake of system improvement. *It would help if, in the future, sponsors of user studies required those who propose them to demonstrate how their descriptive results can or will be used to modify the system described.*

USE

Use may be defined as the extraction of content from a message to meet a need. Orr (1970:155), in his study of needs of biologists, identified what appear to be general needs of scientists and technologists:

I. Regular needs (occur continuously or very frequently)
   A. Current awareness—to keep abreast of new developments.
   B. Everyday reference—to obtain specific items of information or data required for ongoing work or a current decision (e.g., the melting point of a compound, standard specifications, etc.).
   C. Personal
      1. Stimulation—to suggest new ideas and approaches.
      2. Feedback—to obtain reactions to one's own work and decisions that will help to refine problem definitions and solutions and improve quality of work and decisions.
II. Episodic needs (occur irregularly and less frequently)
   A. Retrospective search—to learn of past work relevant to and possibly useful for current or prospective work and decisions.
      1. Exhaustive—"all" relevant work.
      2. Limited—sample of relevant work limited by size or criterion other than relevance (e.g., recency, technical level, authoritativeness, etc.).
   B. Instruction—to acquire new competencies or to brush up in areas where competency has declined.
   C. Consultation—to obtain tailor-made solutions to, or expert opinion on, specific problems.

The requirement for promptness of response varies according to the particular need and work in which the scientist or technologist is engaged.

Identification of needs and uses has been the focus of numerous studies. This is evidenced by the lengthy bibliographies of use studies prepared by Davis and Bailey (1964), which contains 438 references. DeWeese's bibliography (1967) lists 547. Since 1966, the *Annual Review of Information Science and Technology* has devoted chapters in seven of its eight issues to reviews of use studies. Other major reviews include those by Fishenden (1965) and D. N. Wood (1971).

Need studies do not deal with the content of messages the user needs; rather, they are concerned with document-acquisition needs. Brittain (1970:48) observed: "User studies have typically neglected the user as the ultimate processor of information." He continued, "Questions concerning the nature of social science research and the psychology of the user should be examined." Lin and Garvey (1972:10), in their review of the 1971 literature, suggest that "More systematic efforts are needed . . . to isolate factors generating different information needs in terms of their substance and/or source."

Some studies have been conducted to identify sources and channels of information used to meet needs. However, as Lin and Garvey (1972:10-11) note, "Research into types of needs has been uneven, and while many studies have investigated the needs for channels . . . few have focused on the needs for substance (the nature of the material)."

Use Studies

A large portion of the total time spent reading is spent in undirected browsing. Martin and Ackoff (1963) found that chemists spend approximately two-thirds of their reading time in undirected browsing, and the physicist about two-fifths. What they read varies depending on whether they read for specific information or browse (see Table 7-13).

Whereas bibliographies are used most heavily for locating references for primary research, Line (1971) found that use of abstracts and indexes is the most popular means of maintaining current awareness. Their use is followed by personal communications, browsing in bookshops, and consulting book reviews. Informal communications are also used to keep track of work in progress.

Items needed for everyday use are often obtained from reference books. Repetitive needs are usually satisfied by use of personally owned reference books, while less common needs often require the use of a library or other outside services.

For personal stimulation, the user may turn to the literature, discussions with colleagues, students, and conferences. However, Line (1971) found that the user's own research was his greatest stimulus. Conventions, as previously noted, are the first major arena for dissemination of work, and they bring together a major portion of the work conducted in the field

Table 7-13.

Relationship Between What Is Read to Reason for Reading

Part of Journal	Acquired Journal for Specific Information			Acquired Journal to Browse		
	% Chem. Time	% Phys. Time	% Aver. Time	% Chem. Time	% Phys. Time	% Aver. Time
Cover (Front)	0.0	0.0	0.0	6.3	0.0	3.4
Table of Contents	1.0	1.7	1.5	6.9	8.0	7.4
News or Notices	0.0	0.0	0.0	6.8	3.1	5.1
Advertisement	3.7	2.6	3.0	10.1	4.3	7.4
Published Letter	0.0	5.2	3.6	1.6	13.5	7.1
Digest or Abstract	5.7	3.4	4.2	8.5	8.4	8.4
Article	75.0	71.6	74.5	49.6	45.1	47.2
Charts, Diagrams, Pictures	8.6	11.6	10.7	2.6	4.9	3.7
Review (Book, Article)	0.0	1.7	1.2	1.0	3.5	2.1
Index	0.0	0.0	0.0	2.8	0.0	0.9
Other	6.0	2.2	3.3	10.4	1.2	7.3

Source: Miles W. Martin, Jr. and Russell L. Ackoff, "The Dissemination and Use of Recorded Scientific Information," *Management Science* 9 (1963):332. Reproduced by permission of The Institute for Management Science.

during the year. Therefore, they serve as a stimulus for some. Both Utterbach, in his study of "protocols of the process leading to fifteen award-winning industrial innovations," and Baker, in his study of a major corporation (both cited in Allen, 1969) found that the literature is most important in the idea-generating stage of research. Discussion, on the other hand, is most important in problem definition when feedback is required.

In order to receive feedback, the scientist may make prepublication reports or send out preprints to friends and colleagues. At conventions, the presenter is the focus of attention and may receive valuable input from the attendees. According to Lin, Garvey, and Nelson (1970:69) 50 to 60 percent of the scientists surveyed reported that meetings had some effect on their work. One out of every five authors reported "substantive changes in their manuscripts" resulting from feedback from prepublication reports.

Lin, Garvey and Nelson (1970) also examined the preprint-distribution process. Of the authors they studied, 56 percent distributed preprints, and 60 percent of those distributing preprints distributed them prior to publication. The average number of preprints distributed per author was eight. Scientists send preprints to colleagues in the field and to those who have requested previous reprints. Lin, Garvey, and Nelson suggest that social scientists use preprints more heavily than either physicists or engineers because the editorial process is harsher in their field and they want critical feedback before submitting the manuscript.

Menzel (1970) found that scientists use the literature primarily in searches for facts and theories. When looking for procedures, techniques, materials, and apparatus, they turn to personal communications. To identify the sources, scientists may use either the primary literature, abstracts and indexes, or bibliographies. When the scientist or technologist knows the source but does not know where to find it, he often has difficulty in finding the place closest to him that has it.

Menzel also found that retrieval services are important as means for "keeping up." Such services lead to the final information source. Here the literature, and books in particular, are important. Local colleagues often provide leads to the literature, but meetings play no great role in this process.

For a "tailor-made" solution to a problem the scientist or technologist may turn to his friends and colleagues. Informal networks are used to synthesize material and to support findings. To obtain expert advice, the scientist or technologist usually turns to a friend who is an expert, or to someone the friend has suggested.

Martin and Ackoff (1963) found that the chemist spends approximately 18 percent of his time (based on 90.2 hours per week) in scientific communication, but he spends relatively little time reading. On the average, he spends less than three percent of his time reading scientific articles.

"As literature is made more available to the individual chemist through company libraries and particularly through circulation to his desk, he spends much more time reading the material." This suggests that he would read more if he were automatically provided with material that is relevant and useful.

Martin and Ackoff also found that abstracts are used more as a substitute for articles than as a guide to them. This finding along with (a) that indicating users spend more time reading when more that is relevant is made easily available to them and (b) that most of what they receive is neither relevant nor useful, points to the need for filtration, condensation, and consolidation of information reaching scientists and technologists. The Federal Council of Science and Technology (1968:27) observed:

Consolidation has two aspects. First, and of great importance to the research worker, is the critical evaluation and ordering of the primary literature. Second is the selection and simplification of this material, functions that are of absolute necessity for managers (who need to have an understanding of a wide range of technical topics), practitioners (who need to adapt new knowledge to very specific practical problems), and teachers and students at all levels of the educational process. Authoritative studies of technical information and communication . . . have strongly recommended that governmental and other bodies concerned with research and development assume more explicit responsibility for consolidation of information, and that the scientific and technical community grant more prestige and recognition to those who do consolidation work.

Wood (1971:20) has also indicated an increasing awareness of the need for consolidated information: "There is evidence that reviews of progress are wanted by scientists and technologists. This is probably the only way they see of keeping up with the ever-expanding literature of their fields."

Interrelated Channels

Several studies have shown that the usefulness of an information source to a user varies with the nature of his work. The scientist is more likely than the technologist to place a high value on current primary publications, such as journals, and secondary services. The technologist relies heavily on oral communications. He reads journals and scans abstracts journals primarily for current awareness. The scientist relies on personal communication for leads to channels that ultimately deliver the message.

The interaction of formal and informal channels has been examined by Graham, Wagner, Gloege, and Zavala (1968:71). They found that published work provides useful reference information, but that "informal contacts are needed to amplify and clarify information." Similarly, published material is supported and messages are verified by informal contacts. They stated: "It is clearly apparent from the contents of interviewees' comments

that formal and informal research communications are mutually dependent and complementary" (1968:73).

Johnston and Gibbons (1970:24-25) made the following observations on
the use by technologists of both formal and informal channels:

Information contributing to the innovation process is obtained equally from
printed matter and by personal contact . . . . It is apparent that different types
of information are obtained from different sources . . . . The innovation process
was most efficient . . . when these two general sources were combined into a
mutually supportive relationship; in particular this occurred when personal
contacts were used for locating relevant information in printed material, and
subsequently for translating this information into a form appropriate for the
problem-solver.

These results are supported by those of Parker (cited in Allen,
1969:20) who found:

it is not so much total information consumed by the individual that predicts his
productivity, but rather, integration within interpersonal communication channels. More productive researchers do not use impersonal channels significantly
more than less productive researchers.

Johnston and Gibbons (1974) suggest that differences of usage may be
attributed to different criteria used in determining the importance of information inputs.

Summary of Usage

Both formal and informal systems are used to meet the information
needs of scientists and technologists. Certain needs are associated with
certain channels, usage of which varies with job function, organizational
affiliation, education, discipline and the user's evaluation of the channels.
In general, it has been found that formal primary, secondary, and tertiary
messages are important for current awareness, stimulation, and factual
and theoretical searches. When feedback or a synthesis of information is
required, the informal network is frequently called upon.

Scientists and technologists spend relatively little of their time reading
the literature. Therefore, filtering, condensation, and consolidation would
increase their exposure to relevant literature.

DEFICIENCIES

Now that the current system has been described, we summarize the
deficiencies that are revealed by a comparison of it with the ideal design.
We consider first the formal system, then the informal system, and finally
the values and procedures that govern their change.

The Formal "System"

In this discussion we continue to use the functions: production, dissemination, acquisition, and use. The deficiencies do not always fit neatly into these categories. Furthermore, the solution to an acquisition defect may be found in improving dissemination. We list deficiencies where they arise rather than where action may best be taken to remove them.

The principal shortcomings in *production* are:

(1) delays that occur between the preparation of manuscripts and their publication;

(2) waste and delay that occur when input for some secondary services are prepared separately from the primary messages;

(3) the lack of, and delays in, state-of-the-art reviews and other syntheses;

(4) redundancy in primary messages that is encouraged by valuing quantity published over quality of publications;

(5) lack of standardization of secondary-information databases;

(6) absence or shortage of useful secondary information concerning messages in minor media (e.g., film);

(7) delays in translation from foreign languages; and

(8) imprecision of estimating the demand for books and the consequent cost to publishers.

The principal deficiencies of *dissemination* are:

(9) the absence of a single source of all secondary information;

(10) infringement of copyright when documents are reproduced without permission;

(11) unnecessary restrictions on the dissemination of potentially useful information (e.g., those imposed by security classification); and

(12) the frequently heavy and inescapable burden of unsolicited information.

The principal deficiencies of *acquistion* are:

(13) users' ignorance of and inability to find out where to obtain a required document;

(14) lack of standardization, hence the need for users continually to relearn procedures as they make use of different libraries and information services; and

(15) wide geographical variations in the quality of library and bookselling services.

There are also shortcomings of the current SCATT System that relate to *use*:

(16)   authors' and publishers' ignorance as to who is using their material and the value they place on it;

(17)   the user's lack of choice of medium through which he can receive the content of a message; and

(18)   the unsatisfactory design of many user terminals (e.g., microform readers).

The Informal "System"

In recent years informal channels of communication have been increasingly subjected to research—for example, the immediate colleagues of scientists and technologists, "gatekeepers," "invisible colleges," those engaged in informal conversation at scientific meetings, and so on. The importance of these channels in providing criticism, filtering, referring, and amplifying is well recognized. Nevertheless, the following deficiencies are apparent:

(19)   it is too difficult for young scientists and technologists to gain entry into an existing invisible college or to form new ones;

(20)   scientific and technical meetings are not as effective as they might be;

(21)   many research and development establishments are not organized to facilitate informal communication within them;

(22)   those who act as valuable switching centers are frequently overloaded; and

(23)   seeing technological artifacts plays an important part in technology transfer but identifying relevant technology to be seen and determining where it can be seen are often time-consuming and difficult.

Values and Processes Governing Change

The current "system" is geared primarily to the producers and disseminators of information rather than to its users. Therefore, there is much more research on production and dissemination than on use. A change of emphasis is needed, but the system does not change easily.

(24)   The system provides services to authors and marketers as well as to the users of information; it is abused by all of them (e.g., redundancy in primary communications and dissemination of unsolicited information). Effective means for reducing such abuse are not generally available.

In many other service systems one finds encouragement of competition and economic self-sufficiency that provides motivation for self-improvement and, through feedback, the information necessary for adaptive control. In the current SCATT System, however, one finds:

(25) little reward for good service;
(26) direct, indirect, and cross subsidies that are not made explicit;
(27) journal subscriptions automatically incorporated in the subscriptions of professional societies; and
(28) little possibility of competition between libraries except in some major cities.

Although available statistics on utilization provide some essential information for management, they are insufficient to guide fundamental changes, such as the introduction of a new service. For the latter, additional research is necessary. One finds, however:

(29) that most research on users' needs is confined to studies of their patterns of acquisition. Little attention has been paid to the ways in which material is used and to the substance and nature of what is needed; and
(30) that in attempting to meet the demand for secondary services, emphasis has been placed on exhaustiveness rather than filtration. Reducing irrelevant information has been sacrificed for increasing relevant information.

Conclusion

Many of the shortcomings summarized above are obvious symptoms of an overall deficiency in the relationships between the current SCATT System's components. There is a lack of cooperation, coordination, integration, and standardization. The idealized design presented in earlier chapters attempts to overcome these and all the other deficiencies cited in this chapter.

8

EPILOGUE
A Tribute to Our Critics

The process by which the design presented in this report was developed involved interaction with a large number of experts. Their reactions to our work were not restricted to technical matters. Many of their comments dealt with very general methodological and philosophical issues raised by our design process and the design it produced. These issues ought to be resolved, but their resolution is not likely to occur until they have been discussed widely.

The very thorough analysis of our design offered by the experts has induced us to attempt a philosophical and methodological "defense" of our undertaking. We organize this defense under four headings:

(1) Idealization,
(2) Dehumanization,
(3) Charges and Usage, and
(4) Feasibility.

IDEALIZATION

One of our advisers raised the following fundamental set of interrelated questions:

At the end of the study we will have what is intended to be an "ideal" SCATT-System design—more properly, a current "best approximation" to such an ideal. What evidence can the study team offer to support this claim? How can the study monitor verify it in a reproducible fashion convincing to others? Indeed, how can there be assurance that what is offered is indeed a SCATT-System design (let alone "ideal")—i.e., that no essential function or component or specification has been omitted?

Our claims for the design are based on (a) our interpretation of what is and is not known in relevant fields, and (b) the reactions of others to our work. Our claim for relative completeness of the design (relative, that is, to the current state of knowledge and understanding of scientific communication and technology transfer) is supported by the fact that the number of comments we have received from others about omissions has decreased with each successive draft of the design. We have no doubt however, that wider exposure of the design will reveal omissions not yet detected. Therefore, we sought completeness not by attempting the im- possible—specifying criteria that could be applied to establish complete- ness in some absolute sense—but by creating a design process that gives considerable assurance that our product will become more and more complete with continuation of the process.

Recall that we do not claim that that our design is ideal, even for us. Rather, what we tried to do was design a system that would be ideal- seeking. Therefore, the failure to capture in our design every conceivable element of an ideal system in all its fullness is not critical, but a failure to design into the system an ability to move effectively toward an ideal property, whatever it may be, would be critical.

What, asked one of our advisers, are appropriate measures for the idealized SCATT System's performance?

One can no more reasonably perform a "centralized" cost-benefit analy- sis of the SCATT System than he could for a national political-economic system. One can audit, forecast, and even control "costs," but what are the "benefits" in a nationwide public facility of such pervasive conse- quence as a scientific information system? Actually, the principal defense of the democratic principle of organization is that it removes the necessity for centralized analysis and evaluation. It makes possible completely decentralized (individualized) analyses and evaluations and it provides a means of aggregating them into a collective evaluation. That few self- styled democracies have decentralized the provision of social services is due to their lack of democracy, not to its failure. We have tried to design a SCATT System that would be as democratic as possible.

Several of our advisers and critics have asked what assumptions about the future are implicit in our design. For example, one asked: "What set of 'World 2000' scenarios is the study team assuming?" Another asked: "Does SCATT suggest the U. S. culture (or social order) of the future? If so, is it good?"

Recall the characteristics of the idealized design process presented in Chapter 1. It is a design of the System we would construct now if we were free to construct any technologically feasible and operationally viable system we wanted. We do not know what our world will be like at the end of the century but we do know what we would like to see happen now. Therefore, the SCATT System was designed for the "here and now." Our design procedure did not require that we either forecast or assume a particular future. We did assume that many relevant changes in society, science, and technology would take place in the future and we attempted to design a system that would be capable of adapting to them effectively. This, in turn, required that we have some idea as to what kind of changes were possible; but it did not require that we have one or more scenarios of possible futures. This is consistent with our belief that the future (the year 2000, for example) depends at least as much on what is done between now and then as it does on what has already taken place. Therefore, predicting the distant future appears to us to be largely an exercise in futility. We are more interested in designing a desirable future and in finding ways of bringing it about; in creating it, not forecasting it.

Our assumption that the future will continue to be determined at least as much by what has not yet been done as by what has been, is in turn, based on another assumption: that there will be at least as much freedom of individual choice as is currently available. We should like to see greater individual freedom but we do not need to assume it because, in the SCATT System we have designed, each individual would be able to "exploit" all the freedom he has.

We do not assume that the National SCATT System we have designed would "fit" all nations equally well. Much of the "inequality," we believe, derives from differences among nations with respect to individual freedom of choice (particularly for scientists and technologists). We have speculated about how other nations might adapt our design for their own use. We believe it would not be difficult for them to do so because major changes would not be required. We are quite sure that any nation that is concerned about scientific and technological information transfer could benefit from using SCATT as a model.

DEHUMANIZATION

We are grateful to our critical readers for their insistence upon exploring so deeply the ramifications of the problem of the dehumanizing effects of technology. It is apparent from the comments we have received that the technological content of our design has overwhelmed some and given them the impression of a dehumanized "mechanistic" system. This type of reaction disturbs us because however inadequate our technical design, we feel quite sure that we have produced a System more sensitive to human values than any we have come upon heretofore.

The user of the SCATT System designed here would be at least as free as he currently is. The current disorganized system is not eliminated by ours and could well continue with little change if users (and other stake-holders) so desired. More important, however, is the fact that the user of our system would have more choice of ways of obtaining information and more relevant information to obtain than he has in the current system. Standardized procedures are provided to him only as options. He is free to design his own.

We are not sure why SCATT should be regarded as an antihumanistic enterprise but we think our critics have helped us by pointing up three characteristics of the System that are subject to this danger. SCATT, they say, charges all users for all services, the rich and the poor alike; second, SCATT requires a very high degree of personal conformity to System needs; third, SCATT as a computer-based system is inevitably mechanistic and constraining. We would like to consider each of these three points in turn.

Charges

Here two things should be noted. First, charging for consumer products rather than making them available at no cost does not reduce the consumer's or producer's freedom of choice so long as he has the resources needed to acquire or produce them. Because some may not have these resources, we have dealt in detail with subsidies to consumers in order to assure their access to the System. Of course we would prefer a society in which there were no shortage of the relevant resources, but we assumed the continuation of one in which such shortages persisted. Should sufficient resources become available to all, our design would need no change.

Second, a system that does not charge does cost even though those who are served by it do not pay the cost. Someone does pay. By making the charges explicit to those who use or support the use of the system we give them an opportunity they do not currently have: to affect the design and operation of the System by their consuming behavior. To increase the users' influence and control over the System in this way, we believe, is to humanize it.

Our current educational system, which serves without cost to those who directly receive its services, is hardly a humane system. We believe it would be made more humane by a voucher system that would enable every student, or his parents, to select the school to be attended.

Conformism

One of our advisers asked: "Does the SCATT System say, in effect, that to get some indeterminate amount of knowledge transfer improvement requires a possibly unacceptable degree of personal as well as institutional conformity to the System's needs?" We think *not* for reasons already considered, but this question will eventually have to be answered by those who use the System. We have tried to provide them with the ability to use their answers to affect the System's design and operations.

Moreover, user control of the System is augmented by the highly participative type of System management we have designed. We have tried to build more democracy into the management of the System than we believe any comparable system currently has. And we believe that democracy is humanizing, not dehumanizing.

To be sure, a number of requirements are imposed by the System, almost all of them on producers for entry of their products into the System. Because of the length of the discussions describing them they give the impression of being more burdensome and constraining than we believe they actually would be. Consider each requirement in turn.

The coding of a document is a very simple task requiring only a few minutes and it gives the author an opportunity to exert a greater influence than he now has over who would use his document and how it would be used, and it would facilitate his receiving feedback from its users. Furthermore, we do not preclude automated coding as an option.

The requirement for an index and abstract is already common in books and articles. Furthermore, we provide automated procedures to those who find these tasks difficult or burdensome. The requirement for two abstracts is not common but is not unheard of. (Many abstracting journals provide their own rather than an author's abstract.) The second abstract could also be obtained automatically in our System if an algorithm, different from that used to provide the first, was used.

We have provided a way for "mechanically" extracting key-word descriptions from indexes. But here too an author has an opportunity to influence use of his document.

The coding and redundancy check are new requirements. Redundancy checks are made now, but usually informally, by many authors. The procedure we have designed yields the documents to be checked with virtually no effort by the author. He must check the documents thus turned up, but it seems reasonable to require him to do so. Without redundancy checks users would often be forced to waste their time, and this is and would be dehumanizing. Certainly there seems to be a net gain for redundancy checks, even to authors, because they generally spend more time as document users than they would in preparing documents they have produced for entry into the System. Finally, we expect them to produce better documents because of the checks.

The Frankenstein Monster

Computer-based systems such as the one we have designed are viewed by many as mechanistic and constraining. This derives in part from one's natural bias against "giant mechanical brains" and formal systems. They smack of "big brother." More importantly, this bias is a consequence of the incredibly bad computer-based systems that most of us have experienced and the dehumanizing effects they have actually had. In too many cases computer-based systems are inflexible, unforgiving, error-prone, and insensitive to human values.

These weaknesses are *not* inherent in computer technology; they are the result of its faulty use. In fact, a strong case can be made that only through the wise application of computer technology can we cater to individual needs and desires within a "mass society." The SCATT System must store and process a great deal of information about a user in order to treat him as a unique individual; for example, *his* changing interests and idiosyncrasies, restrictions he imposes on the type of outputs he receives, his past use of the System, and special pricing arrangements he prefers. Even the most gifted and humane librarian could not give equivalent consideration to such a wide range of individual needs and desires. Paradoxically, the more we want to be treated individually, the more strain we put upon the generalized capabilities of the computer.

We believe the SCATT System would provide a humane instrument of human communication. Its extensive error-detection and correction procedures, coupled with the diagnostic aids to assist the user who is in trouble, would make the System far more tolerant of errors than less sophisticated "hand-operated" systems are. This is especially important in a system such as SCATT that would serve a large number of users who would have varied skills and familiarity with the System. To err is human; to forgive is good design.

One of the major complaints against computer systems is their inflexibility in responding to changing interests and needs. The strong emphasis placed on adaptability within the SCATT System is intended to overcome this weakness. Its design assumes a continuous evolutionary development rather than the implementation of an "ultimate" system that soon loses touch with current needs, often before the system becomes operational. An abundance of feedback mechanisms is designed into the SCATT System to allow users to express their opinions about how the System is performing and how it might be improved. With attention paid to flexibility and feedback, any reasonable initial design would gradually evolve toward an increasingly powerful and useful system. Emphasis on the process of adaptation, rather than on the System's capabilities at any moment of time, is the key to keeping the SCATT System responsive to human needs and desires.

One of our advisers said that SCATT should strive to make all scientific communication and technology transfer as "simple, fast, and complete as it is where there is direct communication between producer and user." One could argue about how simple, fast, and complete certain kinds of direct communication are—for example, that between some professors and their students. Nevertheless, we take this comment in the spirit in which it was offered and agree with it not only because of the effectiveness that direct communication can have, but also because it can be the most humane type of communication.

Our design has created no obstruction to direct communication other than the charge for unsolicited recorded communications, and this is intended to reduce the inclination (not the ability) of one party to impose communication on another. We consider solicitation by mail or over the telephone, for example, to be dehumanizing.

On the other hand, we have greatly increased the ability of those who want to communicate with each other to do so. SCATT's communication network makes it possible for any number of people, however dispersed, to converse with each other in real-time, and to do so easily. We feel this will enhance invisible colleges, which we believe to be one of the most effective means of scientific communication and technology transfer. Our design makes it possible for associates to share automatically much that each does that is relevant to the others. Indeed, we regard retrieval from the SCATT System to be secondary to retrieval from others, friends or Fellows, through the System.

We anticipate that scientists and technologists will continue to find that other people are more useful sources of information than computers if only for the reason that the "right" people can quickly develop an appreciation of the inquirer and his inquiry, while the computer cannot. Our design incorporates this expectation not only with respect to communication between friends and Fellows, but also with SCATT-Center and affiliated library personnel.

CHARGES AND USAGE

We have referred above to the question of equity in the matter of charges for service. Here we revert to the problem in connection with its possible effect upon use of the System.

One of our advisers wrote: "I keep wondering if a scientist or engineer would participate in SCATT any more than he does in the present array of services." Others have suggested that charging for services would reduce such participation. Such questions have occupied a great deal of our time. We have no answer that can completely dispel such doubts. Only the implementation of the System and its subsequent use can do so. Nevertheless we firmly believe the System would be used much more extensively than "the present array of services," and we think that provision for charges would not seriously reduce usage. We repeat, however, the matter of charging is subject to extensive change or even abolition without seriously affecting the operation of the System. We offer the following observations on the effects of changes on usage.

The SCATT System we have designed will (1) provide more services than are currently available, (2) enable users to use the System from their office or building, (3) provide service more rapidly than does the current array, (4) provide the user with more freedom to design his use of the System, and (5) enable him to affect the design and operations of the System. For these reasons we believe he will use it more than the current one. In fact, we can see no reason why he should not other than the cost involved in so doing.

Most scientists and technologists, however, will not incur these costs personally. Their employers would normally cover them much as they currently cover computer, communication, and transportation costs even when they are explicit. Universities, research sponsors of university-based projects, research institutes, and research and development departments of corporations and public agencies can be expected to cover most of the costs. They do now one way or another, but usually indirectly; hence, they are hidden in overhead. The cost of SCATT usage would become an explicit line item in R&D budgets. When these charges are made explicit, cost-benefit analyses can be more effectively carried out both by the individual user and his sponsor.

If SCATT services are perceived as too costly, they will not be used. This will feed back to the System and encourage it to take corrective action. If it cannot do so, then the service involved would be dropped and, hopefully, some other information-service organization would find a way of providing it at a more attractive cost to the user. There are a number of commercial information services available today that charge for their services much as we have specified for SCATT. Many of these services are thriving and are expanding rapidly.

There is a tendency to overlook the fact that, in many cases in which a previously free service is converted into one for which a charge is made, usage increased significantly. To cite one example: the Busch Gardens in Tampa, Florida—a major tourist attraction—converted from "open to the public at no charge" to charging a fairly substantial fee and in doing so increased attendance and income, enabling its owner to build several additional such family entertainment centers in other parts of the country. The same has been true for many zoos in the United States.

There is another fact that tends to be overlooked: services that are free and are used heavily are more often than not misused or abused. This is clearly the case in many corporate computer installations with which we have been involved. Available computer time is a vacuum that a potential user, like nature, abhors. Free computer services tend to be used inefficiently and for trivial tasks. Once the vacuum is filled with such trivia it is virtually impossible to displace it with something that is important and really needs to be done.

Our experience in design of corporate computer-based management information systems has revealed that when explicit charges are made to the user or his organizational unit for his usage, the quality of his usage increases significantly. At first, when such a pricing system replaces a free service, there is usually an initial reduction of usage; trivial applications are withdrawn. Then the quality of the usage and the quantity of the payoff build up and eventually exceed their previous levels. We expect the same would happen in the SCATT System presented here.

Because this System would not replace, but would only supplement, the current system, there would be no reason for their combined usage to decrease. And because implementation of our design would necessarily be incremental, there should be no trauma associated with the transition to it.

FEASIBILITY

Why, we have been asked, have we incorporated into our design certain features that have been tried, evaluated, and found deficient—for example, author abstracting and indexing? Why are we repeating past mistakes?

This matter is so serious that we must confront it squarely and try to answer the questions it raises in a principled way. Our perfectly general answer is that our design focuses on the *whole* SCATT System and not on its features taken separately. We believe that if the System were implemented, its stakeholders would relate to it more as a whole than as an aggregation of independent parts. Implementation of the System would create a new scientific and technological "culture," a new pattern for producing and consuming information. We believe that the attitudes and behavior of its users would differ significantly from those of the typical users of the current system. It is true that the success of the System would require some modification of the behavior of each type of participant in it. But, we believe, it is equally true that some modifications are not only desirable but are necessary so that science and technology may make continuous progress, and that the societies that contain them may improve. Feasibility thus becomes a question of "feasibility for what?"

In their present condition, science and technology are likely to suffer a nervous breakdown induced by information overload. They are even more likely to become increasingly wasteful of resources through unnecessary duplication of effort brought about because of an inability to find relevant information hidden in a morass of poorly organized data. This wastefulness is increasing precisely at a time when resources required to support science and technology are becoming scarcer. Conservation of resources is becoming more and more closely coupled not only with preservation of science and technology, but also with preservation of the quality of life, if not of life itself.

In the currently disorganized information system, users have very little incentive to provide feedback to it, and what they do feed back has little effect on the system's behavior. Neither would be true for the System we have designed.

The inadequacies, failures, or abortions of other systems are *not* necessarily relevant to our design, and the instruments and means they have used are not necessarily condemned. We do not believe that such failures reveal "intrinsic and unchangeable characteristics of human nature," characteristics that prevent some aspects of our design from being implemented or, if implemented, from being successful. We have tried to design a communication environment that will reveal what we believe to be the true nature of such characteristics of human nature: that they are defenses against the abuses of man by the systems that contain him.

To some the System we have designed seems infeasible because of the complexity and variety of its operations. "Why," they have asked, "should there be one system to handle all scientific and technological information?" They suggest that autonomous discipline-, profession-, and problem-based information systems are and will be sufficient to meet our needs; and this would avoid centralization and the authoritarianism that centralization tends to breed. This suggestion assumes either that science, technology, and their environment will not change radically, or that if they do, such change will occur slowly enough to permit the currently unresponsive system to respond adequately.

Let us recall a constraint that we imposed on our design effort: we did not allow ourselves to redesign the environment in which scientific communication and technology transfer takes place. Therefore, we accepted the environment as it is and as it might be. We do not know what it will be, but we do know that science, technology, and society will change and, in all likelihood, more rapidly and fundamentally in the future than they have in the past. Therefore, we attempted to design a System that was flexible and adaptive, capable of operating effectively in a rapidly changing environment. Most of the complexity of our design is a consequence of this effort.

We know that even today social crises arise with increasing intensity at decreasing intervals. Just one instance from the recent past: we have not been able to organize our current disorganized SCATT System so that it would respond rapidly and effectively enough to meet the energy crisis. The informational needs created by social crises do not respect disciplinary, professional, or previously defined problem boundaries. Therefore, there is a need for a unified system whose content can rapidly be reorganized to meet changing requirements.

We can expect an accelerating evolution of science and technology and, therefore, what may well become a virtually continuous "paradigmatic crisis" within them. New disciplines, cross-disciplines, interdisciplines, and metadisciplines will arise with increasing frequency and will have to arise if science and technology are to be responsive to the needs of an increasingly large and complex society. The organization of science and technology and the professions based on them will approach continuous change, thereby requiring continuous changes in the organization of the information they use and produce. This requirement is best met, we believe, by the reorganization of material contained in a single system rather than by an attempt to reorganize the content of many autonomous or semiautonomous systems. The reorganization of a currently simple system is a very complex thing.

The inability of our current information systems to respond rapidly or effectively to the emergence of new scientific and technological paradigms, new fields of study, new professions, and new methodologies is an obstruction to their development. We have tried to remove this obstruction by making the SCATT System flexible and adaptive. This cannot be done without making it more complex. Such complexity is required to make the science-technology system and the society that contains it capable of surviving in a very turbulent environment. In such an environment simplicity and survival become antithetical. "Only the simple is feasible" is a song sung by Sirens who entice science and technology, even if unconsciously, to self-destructive acts. There is at least one kind of simplicity that must be avoided: simple-mindedness.

Let us now consider the feasibility question in a more general context. One adviser wrote: "Every time I found myself concluding that some parts of the idealized SCATT just never could be implemented for legal, vested interests, economic, or practical reasons, the message in Chapter 1 kept me reading on. Nevertheless, there is much about SCATT that strikes me as too idealistic." In contrast, when an earlier version of our design was presented to a meeting of the American Society of Information Science in Atlanta, one member of the audience observed that the characteristic of our design that most struck him was how close to realizable it was. The principal obstruction to its realization, he remarked, was the state of mind of those who assert it is not feasible.

Naturally, we are sympathetic to this second comment but this does not mean that we consider the first to be capricious. Of what value, it asks, is a design whose elements are so largely unrealizable?

Our reply lies in an observation made in Chapter 1: a design, all or most of whose elements when considered separately appear to be infeasible, may nevertheless be feasible or nearly so. This follows from the fact that a system always has properties that none or some of its parts do not have. (Note that the operations of the human heart would not be feasible unless it were considered as part of a body.) A design element that appears infeasible when considered separately may appear feasible when considered as a part of a whole.

In the first comment quoted above, reference is made to "parts . . . [that] just never could be implemented." We agree—*if they are taken separately*. But with our second observer we believe that, taken together, most of them can be. This does not mean that we believe they *will* be. We hope, however, that the design and this report of it will increase the likeli-hood that it will be. The philosopher Ortega y Gasset (1966) reassures us here: "man has been able to grow enthusiastic over his vision of . . . unconvincing enterprises . . . And in the end he has arrived there."

The principal obstruction to the attainment of man's ideals is man him-self. We know of no more effective way of removing this obstruction than by providing man with a vision that induces him to overcome himself. We do not believe that our design of a SCATT System provides such a vision, but we do believe that we have initiated a process that can produce such a vision.

REFERENCES

Ackoff, Russell L. *Choice, Communication, and Conflict*. Philadelphia: Management Science Center, University of Pennsylvania, 1967.
The content of this report is incorporated to a great extent in Part III of Ackoff and Emery (1972).

———. *A Concept of Corporate Planning*. New York: John Wiley & Sons, 1970.
———. *Redesigning the Future*. New York: John Wiley & Sons, 1974.
——— and Fred E. Emery. *On Purposeful Systems*. Chicago: Aldine-Atherton, 1972.
Book describing the fundamentals of their conceptual system of behavioral theory. Particularly relevant to this report is Part III, "Interactions of Purposeful Systems" in which communication is discussed.

Allen, Thomas J. "Performance of Information Channels in the Transfer of Technology." *Industrial Management Review* 8 (1966):87-98.
Study of performance of channels for transferring technical information, indicating differences in their impact on research.

———. "Information Needs and Uses." In Cuadra and Luke (1969), pp. 3-29.
———. "Communication Networks in R & D Laboratories" *R & D Management* 1 (1970):14-21.
Study of the impact of physical location and project organization on communication, coordination, and informal relations in an R & D lab.

——— and Alan R. Fusfeld. "Research Laboratory Structuring of Communication." Working Paper, January 1974, #692-74.
Paper reviewing studies of communication in R & D organizations that show the influence of physical arrangement on communication. It describes an experiment to improve communication through an architectural change.

——— and Peter G. Gerstberger. "Criteria Used by Research and Development

Engineers in the Selection of an Information Source." *Journal of Applied Psychology* 52 (1968):272-279.

———, J. M. Piepmeir, and S. Cooney. "The International Technology Gatekeeper." *Chemical Industry Developments Incorporating CP & E* (1972):35-41.
International technology transfer through a gatekeeper is proposed as a way for a small nation to keep abreast of developments.

American Psychological Association (APA). *Reports on the Project on Scientific Information Exchange in Psychology.* Vols. 1-3. Washington, D.C.: American Psychological Association, 1963-69.
"an intensive and comprehensive study of the scientist's information-exchange behavior within his scientific environment" (1965, p. 1).

Auerbach Corporation. *DoD User Needs Study.* Philadelphia, Pa., 1965.
Examination of the use by DoD scientists of formal and informal sources of information.

Bach, Harry B. "A Comparison of *Operations Research* and *Management Science* Based on Bibliographic Citations." *INTERFACES* 4 (1974): 42-52.
Study of the similarities and differences between these journals through an analysis of their citation patterns.

Barnes, R. C. M. "Information Use Studies Part 2—Comparison of Some Recent Studies." *Journal of Documentation* 2 (1965):169-176.
A survey of user studies and the problems limiting comparison of the results.

Brittain, J. Michael. *Information and Its Users.* New York: Wiley-Interscience, 1970.
Comprehensive survey and review of user studies in the social sciences and the methodology used. It also indicates the neglected aspects of users and their information requirements.

Brodman, Estelle. "Choosing Psychology Journals." *Medical Library Associations Bulletin* 22 (1944):479-483.
Study comparing the ranking of periodicals by members of the Department of Physiology of Columbia University with a ranking by number of citations.

Chen, Ching-Cheh. "Use Patterns of Physics Journals in a Large Academic Research Library." *Journal of ASIS* 23 (1972):254-265.

Crane, Diana. *Invisible Colleges: Diffusion of Knowledge in Scientific Communities.* Chicago: University of Chicago Press, 1972.
A work on invisible colleges, networks, and the diffusion of information.

Crawford, Susan. "Informal Communication Among Scientists in Sleep Research." *Journal of ASIS* 22 (1971):301-310.
Study of the informal network of sleep scientists that identifies core members.

Cuadra, Carlos A. and Ann W. Luke. *Annual Review of Information Science and Technology.* Vols 4-9. Washington, D.C.: American Society for Information Science, 1969-1973.
Comprehensive annual review of developments in information science. References here to the reviews included in it are not annotated.

Davis, R. A. and C. A. Bailey. *Bibliography of Use Studies*. Philadelphia: Drexel Institute of Technology, Graduate School of Library Science, 1964.
438 references to user studies.

DeWeese, L. C. "A Bibliography of Library Use Studies." In *Report on a Statistical Study of Book Use,* edited by A. K. Jain, BB 176525 CFSTI, 1967.

Federal Council of Science and Technology. *Making Technical Information More Useful*. Washington, D. C.: Office of Science and Technology, 1968.

Fenichel, C. (ed). *Changing Patterns in Information Retrieval*. Washington, D.C.: ASIS, 1974.
A collection of what are regarded as key papers in the area.

Fishenden, R. M. "Information Use Studies Part I—Past Results and Future Needs." *Journal of Documentation* 21 (1965):163-168.
Brief survey of conclusions that pertain to the development of information services; it indicates areas in which further research would be useful.

Freeman, James E., James P. Kottenstette, and Martin D. Robbins. *New Information Services in Social Problem Areas*. University of Denver: Denver Research Institute, Industrial Economics Division, August 1973.

Freidas, Thelma. *Literature and Bibliography of the Social Sciences*. Los Angeles: Melville Publishing Company, 1973.

Frycki, Stephen J. "Information Transfer from Source to User Utilizing a Pharmaceutical Data Base." In Keenan (1973), pp. 83-98.
Describes the development of an information storage and retrieval system for Squibb Institute of Medical Research.

Gannett, Elwood K. "Primary Publication Systems and Services." In Cuadra and Luke (1973), pp. 243-276.

Garfield, Eugene. "Citation Analysis as a Tool in Journal Evaluation." *Science* 178 (1972), pp. 471-479.
Summary of the results of the analysis of journal citation patterns across science and technology by the Institute for Scientific Information (ISI).

Garvey, William D. and Belver C. Griffith. "Communication and Information Processing Within Scientific Disciplines: Empirical Findings for Psychology." *Information Storage and Retrieval* 8 (1972):123-136.

Garvey, William D., Nan Lin, and Carnot E. Nelson. "Some Comparisons of Communication Activities in the Physical and Social Sciences." In Nelson and Pollock (1970), pp. 61-84.

———. "Information Exchange Associated with National Scientific Meetings in Relation to the General Process of Communication in Science." In *The Role of the National Meeting in Scientific and Technical Communication,* Center for Research in Scientific Communication. John Hopkins University, Vol. 1, June 1970.
Detailed report of the authors' studies on the national meeting.

——— and Kazvo Tomita. "Research Studies in Patterns of Scientific Communication: II The Role of the National Meeting in Society and Techni-

cal Communication." *Information Storage and Retrieval* 8 (1972): 159-169.
Second of a four-part series on the studies conducted between 1966 and 1971 on the information exchange of scientists and engineers from nine physical, social, and engineering sciences. This article presents the authors' studies on the national meeting as part of the overall process of dissemination of scientific and technical information.

Garvey, William D., Nan Lin, and Kazvo Tomita. "Research Studies in Patterns of Scientific Communication: III Information-Exchange Process Associated with the Production of Journal Articles." *Information Storage and Retrieval* 8 (1972), pp. 207-221.
Third part in the series described above; it focuses on prepublication exchange and the assimilation and use of the information by scientists.

Gechman, Marvin C. "Machine-Readable Bibliographic Data Bases." In Cuadra and Luke (1972), pp. 323-378.

Graham, W. R., C. B. Wagner, W. P. Gloege, and A. Zavala. *Exploration of Oral/Informal Technical Communications Behavior.* Clearinghouse for Federal Scientific and Technical Information, AD 669 586, 1967.

Greenberger, Martin, Julius Aronofsky, James L. McKenney, and William Massy. *Networks for Research and Education.* Boston, Mass: MIT Press, 1974.
Summary of the state-of-the-art of networks; it contains papers and discussions from three seminars conducted by EDUCOM with the support of NSF in late 1972 and early 1973.

Griffith, Belver and A. James Miller. "Networks of Informal Communication Among Scientifically Productive Scientists." In Nelson and Pollock (1970), pp. 125-140.

Gruber, William H. and Donald G. Marquis. *Factors in the Transfer of Technology.* Cambridge, Mass.: MIT Press, 1969.

Halbert, Michael H. and Russell L. Ackoff. "An Operations Research Study of the Dissemination of Scientific Information." In *International Conference on Scientific Information,* pp. 87-120. Washington, D. C.: National Academy of Science, National Research Council, 1958.
Report on the authors' research on how scientists spend their time, with particular emphasis on their allocation of time for communication.

Havelock, Ronald G. *A National Problem-Solving System: Highway Safety Researchers and Decision Makers.* University of Michigan: Institute for Social Research, Center for Research on Utilization of Scientific Knowledge, 1971.

Hawkins, Donald T. "Bibliographic Data Base Usage in a Large Technical Community." *Journal of ASIS* 25 (1974):105-108.

Hayes, Robert M. "Bibliographic Processing and Information Retrieval." In Greenberger, et al. (1974), pp. 161-164.
Discussion of the possibility of linking existing on-line catalogues and reference databases into a National Science Computer Network.

——— and Joseph Becker. "Mechanized Information Services." In Keenan (1973), pp. 99-128.
Presentation of the concept of a Center for Information Services, and the possible problems that would be encountered in setting one up.

Herschman, Arthur. "A Program for a National Information System for Physics." In Nelson and Pollock (1970), pp. 307-324.

Holm, Bart E. "Library and Information Center Management." In Cuadra and Luke (1970), pp. 353-376.

Houseman, Edward M. "Selective Dissemination of Information." In Cuadra and Luke (1973), pp. 221-242.

Institute for Scientific Information (ISI). Promotional literature for *Current Contents*.

Johnston, Ron and Michael Gibbons. "Characteristics of Information Usage in Technological Innovation." Manchester, U.K.: Liberal Studies in Science, May 1974. (Mimeo.)
Report on the results of a study of the characteristics of information used in the process of solving technical problems and contributing to innovation.

Keenan, Stella. *Key Papers on the Use of Computer-Based Bibliographic Services.* Philadelphia: American Society for Information Science (ASIS) and National Federation of Abstracting and Indexing Services (NFAIS), 1973.
Collection of papers written between 1969 and 1973.

Kochen, M. and R. Tagaliacozzo. "Matching Authors and Readers of Scientific Papers." *Information Storage and Retrieval* 10 (1974):197-210.

Krauz, T. K. and C. Hillinger. "Citation, References, and Growth of Scientific Literature: A Model of Dynamic Interaction." *Journal of ASIS* 22 (1971):333-336.

Leimkuhler, Ferdinand F. "The Bradford Distribution." Journal of *Documentation* 23 (1967):197-207.
Summary of Bradford's law of scatter and distribution, together with recent developments using computer-aided retrieval.

Levin, K. D. and H. L. Morgan. "Optimizing Distributed Data Bases: A Framework for Research." *Proceedings of the 1975 National Computer Conference,* May 1975.

Lin, Nan and William D. Garvey. "Information Needs and Uses." In Cuadra and Luke (1972), pp. 5-38.

———, and Carnot E. Nelson. "A Study of the Communication Structure of Science." In Nelson and Pollock (1970), pp. 23-60.

Line, Maurice B. "Information Uses and Needs of Social Scientists: An Overview of INFROSS." *Aslib Proceedings* 23 (1971):412-434.
Report on the study begun in 1967 to examine the information use of social scientists in Britain. This report covers the first phase, the identification of information needs and uses of social scientists.

Lipetz, Ben-Ami. "Information Needs and Uses." In Cuadra and Luke (1970), pp. 3-31.

Martin, Miles W., Jr. "The Measurement of Value of Scientific Information." In

Operations Research in Research and Development, edited by B. V. Dean, pp. 97-123. New York: John Wiley & Sons, 1963.

—— and Russell L. Ackoff. "The Dissemination and Use of Recorded Scientific Information." *Management Science* 9 (1963):322-336.
Report on results of a study on how chemists and physicists spend their time, dealing in particular with the amount and kinds of communication they engage in.

Martino, J. P. "Citation Indexing for Research and Development Management." *IEEE Transactions of Engineering Management* 18 (1971):146-151.
Review of some studies on the utility of a citation index for research.

McCarn, Davis B. and Joseph Leiter. "On-Line Service in Medicine and Beyond." *Science* 181 (1973):318-324.
Description of the National Library of Medicine's (NLM's) on-line bibliographic retrieval system, MEDLINE.

Menzel, Herbert. *Formal and Informal Satisfaction of the Information Requirements of Chemists.* New York: Columbia University and New York University, June 1970.
Summary of a study on the relationship between the users of information channels and the nature of the need; the different uses of formal and informal channels of 161 polymer chemists were studied in particular.

Minsky, M. (ed). *Semantic Information Processing.* Cambridge, Mass.: MIT Press, 1968.
A selection of articles on "artificial intelligence" and computer programs "understanding" natural language. The editor recognizes the limitations in applications of existing results.

National Federation of Abstracting and Indexing Services (NFAIS). *Member Service Descriptions.* Report No. 6. Philadelphia: NFAIS, 1973.

——. *Member Service Statistics 1957-1974.* Philadelphia: NFAIS, 1974.

Nelson, C. D. and D. K. Pollock. *Communication Among Scientists and Engineers.* Lexington, Mass.: D. C. Heath and Company, 1970.
Collection of articles on information needs and uses.

New York Times. "A Surfeit of Journals." (June 29, 1975).

Olson, Edwin E. and Russell Shank. "Library and Information Networks." In Cuadra and Luke (1972), pp. 279-322.

Organization for Economic Cooperation and Development (OECD). *Information for a Changing Society.* Paris: OECD, 1971.

Orr, Richard H. "The Scientist as an Information Processor." In Nelson and Pollock (1970), pp. 143-190.
Summary of the first phase of a project for the Committee on Biological Information to improve A/I services. It discusses trends in S/T policies regarding STI, and future changes in primary publications and libraries.

—— and Alice A. Leeds. "Biomedical Literature: Volume, Growth and Other Characteristics." *Federation Proceedings* 23 (1964):1310-1331.
Overview of a broad study of characteristics of biomedical literature which affect communication.

Ortega y Gasset, Jose. *Mission of the University.* New York: W. W. Norton & Company, 1966.

Park, Margaret K. "Computer-Based Bibliographic Retrieval Services: The View from the Center." In Keenan (1973), pp. 139-142.
Brief discussion of aspects of the relationship between the Information Center and its users and suppliers.

Price, D. J. deSolla. "The Structures of Publications in Science and Technology." In Gruber and Marquis (1969), pp. 91-104.

―――. *Little Science, Big Science.* New York: Columbia University Press, 1963.
A short but much cited book analyzing the trends towards 'Big Science.'

Prywes, Noah S. and Diane P. Smith. "Organization of Information." In Cuadra and Luke (1972), pp. 103-158.

Roberts, Lawrence G. "Data by Packet." *IEEE Spectrum* 11, 2 (Feb. 1972): 46-51.

Rosenbloom, Richard S. and Francis W. Wolek. *Technology and Information Transfer.* Boston, Mass.: Harvard Business School, 1970.
Report on a study of how scientists and engineers acquire information and the relationship of the information sources to the purposes of the search.

Salton, G. *Automatic Information, Organization, and Retrieval.* New York: McGraw-Hill, 1968.
Detailed analysis of computerized document retrieval. The author is responsible for the experimental SMART System.

――― (ed). *The SMART Retrieval System. Experiments in Automatic Document Processing.* Englewood Cliffs, N.J.: Prentice Hall, 1971.
A collection of papers on the SMART System.

SATCOM (Committee of Scientific and Technical Communication of the National Academy of Sciences–National Academy of Engineers). *Scientific and Technical Communication.* Washington, D. C.: National Academy of Sciences, 1969.
Report on a three-year survey of scientific and technical communication. It includes recommendations for increasing the effectiveness of scientific and technical information.

Shannon, C. E. and W. Weaver. *The Mathematical Theory of Communication.* Urbana: University of Illinois Press, 1949.

Small, Henry. "Co-Citation in the Scientific Literature: A New Measure of the Relationship Between Two Documents." *Journal of ASIS* 24 (1973): 265-269.
Brief discussion of this new form of analysis and possible applications.

――― and Belver Griffith. "The Structure of Scientific Literature 1: Identification and Graphing Specialties." *Science Studies* (1974), pp. 17-40.
Report of an experiment to study the structure of a scientific specialty using co-citations.

U.S. Bureau of the Census. *Statistical Abstract of the United States 1973* (94th Edition). Washington, D.C., 1973.

Urquhart, D. J. "The Distribution and Use of Scientific and Technical Information." *Journal of Documentation* 3 (1948):222-231.

Vallee, Jack. *ARPA Policy—Formulation Interrogation Network (FORUM), Final Report.* Menlo Park, California: Institute for the Future, 1974.

Vinken, Pierre J., MD. "Developments in Scientific Documentation in the Long Term." *Journal of ASIS* 25 (1974):109-112.
Opinion paper on the data-bank of the future.

Vinsonhaler, John F. and Robert D. Moon. "Information Systems Applications in Education." In Cuadra and Luke (1973), pp. 277-318.

Virgo, Julie A. "The Review Article: Its Characteristics and Problems." *The Library Quarterly* 41 (October 1971):275-291.

Wade, Nicholas. "Citation Analysis: A New Tool for Science Administrators." *Science* 188 (2 May 1975):429-433.
Discussion of potential uses of citation analysis.

Weinstock, Melvin. "Abstracting and Indexing Services." In *Encyclopedia of Computer Science and Technology,* edited by Jack Belzer, pp. 142-166. New York: Marcel Dekker, 1975.

Weiss, Stanley D. "Management Information Systems." In Cuadra and Luke (1970), pp. 299-324.

Withington, Frederick G. "Beyond 1984: A Technology Forecast." *Datamation* (January 1975):54-73.

Wood, David N. "Discovering the User and His Information Needs." *Aslib Proceedings* 21 (1969):262-270.

———. "User Studies: Review 1966-1970." *Aslib Proceedings* 21 (1971):11-23.
Review which updates Fishenden's (1965) article.